Lecture Notes in Engineering

Edited by C. A. Brebbia and S. A. Orszag

71

C. M. Lemos

Wave Breaking

A Numerical Study

Springer-Verlag
Berlin Heidelberg New York
London Paris Tokyo
Hong Kong Barcelona Budapest

Author
Carlos M. Lemos
Dept. of Coastal and Estuarine Dynamics
Instituto Hidrografico
Rua das Trinas 49
1296 Lisbon Codex
Portugal

ISBN-13:978-3-540-54942-0 e-ISBN-13:978-3-642-84688-5
DOI: 10.1007/978-3-642-84688-5

Typesetting: Camera ready by author
61/3020-5 4 3 2 1 0 Printed on acid-free paper.

To my wife São and my daughter Mariana

PREFACE

In this work a numerical model for studying two-dimensional wave motion in the vertical plane is developed. The basic elements of the present methodology are the time-dependent Reynolds equations for incompressible flows with Boussinesq (ν_t) closure, a especially developed K-ϵ turbulence model for describing the turbulent transport processes, and an efficient technique (VOF—Volume Of Fluid) for defining and updating the free-surface. This approach can potentially describe the propagation, deformation and overturning of pre-breaking nearly irrotational waves; and the free-surface configuration, mean velocity and pressure, and turbulence field in broken waves. This technique unifies the study of wave transformation processes in shallow water, which so far required the use of several different approaches, ranging from analytical to empirical and experimental. The numerical model was applied to several wave transformation problems, demonstrating its predictive capabilities for a joint description of wave motion before, during and after breaking.

I am grateful to the Portuguese Navy and to the Junta Nacional de Investigação Científica e Tecnológica (Lisbon, Portugal) for sponsoring this investigation. Institutional support from the Instituto Hidrográfico and from the Portuguese Army Charting Service (Lisbon, Portugal) is also gratefully acknowledged. The numerical model here presented was developed in the Department of Hydraulic, Maritime and Environmental Engineering of the Catalonia University of Technology (Barcelona, Spain). Thanks are due to Professor A. S. Arcilla for his help and interest in my work and to Mr. F. Collado for his valuable suggestions. I am particularly indebted to Dr. D. A. Rodrigues of the University of Lisbon (Portugal) for his helpful and stimulating discussions. Finally, I wish to express my gratitude towards all members of my family, especially towards my wife, for their continuous and unfailing help, love, and trust in my achievements.

Lisbon Carlos M. Lemos

May 1991

TABLE OF CONTENTS

Chapter 1

INTRODUCTION

1.1 Nature and scope of the work.

Wave transformation phenomena in shallow water are of great theoretical and practical importance. From the viewpoint of coastal engineering, nearshore waves act as an external forcing to coastal structures, and as the main driving agent for nearshore currents and sediment transport. From the viewpoint of theoretical fluid dynamics, wave transformation processes produce some important (but unsolved) problems on water wave dynamics.

Pre-breaking wave motion is nearly irrotational. Therefore, several fully theoretical approaches are available for estimating wave evolution (Peregrine, 1983; Nadaoka, 1986). Wave transformation prior to breaking is characterized by the effects of nonlinearity and dispersivity of waves, and by the effect of disturbances (change of water depth, other waves, wind, etc). The waves maintain their form if nonlinear steepening and dispersion are in balance. Disturbances (mainly the change of water depth) break the equilibrium, induce instabilities, and cause overturning and breaking. After breaking, intense vorticity and turbulence are generated by the folding of the free-surface onto itself. The wave´s dynamics change abruptly, owing to the complicated structure of the velocity field and the high energy dissipation and diffusion. The complexity of turbulent transport processes precludes the application of the deductive approaches used for pre-breaking (irrotational) waves. Consequently, most investigations on the details of broken wave dynamics have been experimental (Mizuguchi, 1986; Nadaoka, 1986).

The study of breaking wave propagation and decay can be carried on by using two types of methodologies: by a description of the details of wave motion, or by a description of the wave field in terms of the spatial distribution of their integral properties. The global description is the basis of most nearshore circulation models, and has the advantages of economy and simplicity. However, some basic assumptions of global models are not valid inside the surf-zone. Thus, the quality of the global approach can be greatly improved through a better understanding of the details of breaking wave motion.

This work is concerned with the former type of methodology, dealing with phenomena that take place in time scales comparable to a wave period. A numerical approach for studying two-dimensional wave motion in the vertical plane is developed. This approach is potentially able to describe propagation, deformation, and overturning of pre-breaking waves; and the free-surface configuration, mean velocity and pressure, and turbulence field in broken waves. However, it also has real limitations: the effects of air entrainment (non-negligible compressibility and density variations) are neglected, and some aspects of turbulent transport (*e.g.* redistribution, anisotropic diffusion and dissipation, etc.) are not described.

1.2 Methodology.

The numerical simulation of turbulent free-surface flows undergoing large deformations is an important challenge in computational fluid dynamics. The main difficulties are the treatment of multiple free boundaries and the adequate representation of turbulence dynamics. The numerical approach here developed is based on the following elements:

- Numerical integration of the time dependent incompressible Reynolds equations with the Boussinesq (eddy-viscosity) closure assumption.

- Free-surfaces are defined and updated using the volume-of-fluid (VOF) technique (Nichols & Hirt, 1975; Nichols et al. , 1980).

- Turbulent transport processes are described by means of a especially developed K-ϵ turbulence model.

The Reynolds equations were used instead of the Navier-Stokes equations because of the practical impossibility of calculating the turbulent fluctuations. The volume-of-fluid (VOF) technique for treating multiple free-boundaries was chosen because of its accuracy and computational efficiency. The K-ϵ, 1-point closure, 2-equation turbulence model was chosen because it is capable of a moderate description of the turbulent transport processes, has been applied in many engineering and industrial flows, and represents a good balance between cost, accuracy and effectiveness (Rodi, 1980).

In what follows the contents of the present work will be described.

In *chapter* 2, the basic theory of incompressible fluid dynamics is reviewed, with emphasis on free-surfaces, pressure solution, turbulence dynamics and initial/boundary conditions. Turbulent transport processes are discussed in terms of transport equations for turbulence-related variables. A turbulence model is selected by taking into account the physical characteristics of surf-zone turbulence and cost *vs* effectiveness considerations.

In *chapter* 3, the physical processes of surf-zone wave transformation are analyzed. The limitations of the existing mathematical descriptions are discussed in a comparative manner, and the elements of the present methodology are outlined.

In *chapter* 4, finite difference (MAC-type) methods for incompressible fluid dynamics are reviewed. Different methods for pressure solution and free-surface representation are discussed, and the basic structure of the solution algorithm is selected.

In *chapter* 5, the numerical model is described. The discretized equations and initial/boundary conditions are presented, as well as the appropriate stability criteria. The structure of the algorithm is analyzed in detail. Finally, some preliminary tests are done, using several types of flow (laminar, turbulent, confined and with free-surfaces) for which the solution is known.

In *chapter* 6, the numerical model is applied to shallow water wave transformation problems. The sequence *disturbance* ⇒ *wave evolution* ⇒ *breaking*, and the mean motion and turbulence fields after breaking are computed for moving hydraulic jumps, solitary waves, and periodic waves. Global wave properties are not evaluated, for this would require phase averaging over a large number of periodic waves.

In *chapter* 7, the most important conclusions are summarized, and suggestions for future research and development are presented.

1.3 Innovations and conclusions.

The main contribution for the study of water wave dynamics, was the introduction of a numerical model that is capable of representing the wave transformation processes before, during and after breaking in a unified way. The most relevant contributions for computational fluid dynamics were :

- The K-ϵ turbulence model was combined with an efficient algorithm for solving the incompressible Navier-Stokes equations with arbitrary free-surface configuration, defined using the VOF technique.

- Simplified boundary conditions for the K-ϵ model in breaking free-surface flows were formulated and discussed.

- A heuristic (objective analysis) method for specifying initial conditions for K and ϵ in the absence of suitable analytical and experimental solutions was introduced and discussed.

- Semi-implicit schemes for treating the production and dissipation terms in the modeled K-ϵ equations were introduced and compared.

- A heuristic stability criterion for the production and dissipation terms in the modeled K-ϵ equations was introduced. This criterion relates the physical problem of breakdown of the basic assumptions of the K-ϵ model, with the numerical problem of stability.

The most important conclusions for computational fluid dynamics were :

• The present model conserves mass and momentum with high accuracy. These conservation properties are not affected by the turbulence submodel.

• The modified FTCS scheme proved to be a simple and valuable discretization for convection-dominated flows.

• Spurious interactions between the components of the model were not observed.

• The computer program was tested with a large number of different flow problems, and proved to be an extremely flexible tool for scientific and engineering applications.

• The numerical model developed was optimized for vector and parallel processing. This feature will be important in the future, in view of the increasing number of institutions having access to supercomputers.

The relevant conclusions for the hydrodynamics of shallow water waves were:

• The computed critical depth ratio for the hydraulic jump was found to agree with the experimental value, in contrast with previous numerical calculations based on Boussinesq-type shallow water equations. This discrepancy is due to the fact that the velocity field in transitional jumps cannot be described using low-order shallow water approximations.

• It was found that hydraulic jumps and solitary waves start breaking by the formation of a single convergent plunging jet, such as occurs in other types of waves (*e.g.* deep- and shallow-water periodic waves). This reinforces the theoretical interest of solitary waves and hydraulic jumps.

• The order of approximation of the wave theory used for generating initial and boundary conditions was found to have a great influence in the quality of the numerical solution for highly nonlinear waves.

• Two numerical experiments on the transition from pre-breaking to broken wave motion were performed, one using a solitary wave, the other using periodic waves. All processes of wave transformation across a surf-zone — deformation, overturning, production and dissipation of vorticity and turbulence, transformation into a moving jump — were correctly represented. The results showed that in both cases the region with high turbulence intensity, eddy-viscosity and shear stress was found above the trough level. The computed distributions of the turbulence variables were more concentrated, and had higher maxima, than those found in physical experiments under similar conditions. This discrepancy is due to limitations of the present K-ϵ model. However, it should be noted that the experimental techniques also have important limitations.

Chapter 2

GENERAL ASPECTS OF INCOMPRESSIBLE FLOW THEORETICAL REVIEW

2.1 Introduction.

This chapter presents a review of the basic equations governing incompressible flow, with special emphasis on the representation of free-surfaces and turbulence dynamics. To simplify the mathematical treatment, certain restrictions are introduced. The fluid is assumed incompressible, Newtonian, and with uniform density.

The incompressibility/uniform density hypothesis is violated in zones with strong turbulence close to the free-surface in breaking flows (the so-called "roller"). It is a reasonable approximation because air is much less dense than water, and because a detailed description of the two-phase mixture in a roller is presently beyond the state-of-the-art (due to a lack of precisely formulated governing equations, state and constitutive equations, etc). The effects of sediment transport and percolation in the dynamics of water motion are also neglected.

The outline of this chapter is as follows. In §§ 2.2—2.6, the basic theory of the Navier-Stokes equations is reviewed. The problems posed by free-surfaces are considered in detail. The role of the pressure Poisson equation is analyzed together with the problem of pressure boundary conditions. The significance of the energy and vorticity equations for the interpretation of turbulence dynamics is introduced in §§ 2.4—2.5. In §§ 2.7—2.10, the turbulence closure problem is considered. After introducing some basic aspects of turbulent flows (§ 2.7), the dynamics of turbulent transport are characterized in terms of transport equations for some turbulence-related variables (§ 2.8). Examination of realistic closure schemes for two of these equations (the so-called K and ϵ equations) leads to the formulation of a turbulence model that is consistent with the physical characteristics of breaking-flow turbulence, yet relatively simple to implement.

2.2 The Navier-Stokes equations for uniform, incompressible fluids.

The motion of an incompressible, uniform density fluid is governed by the continuity and momentum equations. These are the mathematical expression of the mass and momentum conservation principles, respectively .

The continuity equation is derived by assuming that the mass of a fluid particle remains constant. If δm is the mass of the particle and δV its volume, the logarithmic derivative of δm is

$$\frac{1}{\rho}\frac{D\rho}{Dt}+\frac{1}{\delta V}\frac{D(\delta V)}{Dt}=0$$

using D/Dt to denote the material derivative, $D/Dt \equiv \partial/\partial t+\vec{u}\cdot\nabla$. The rate of volume dilatation $[1/\delta V]\cdot[D(\delta V)/Dt]$ is equal to the divergence of velocity . Hence, if the fluid is incompressible and the density uniform, the continuity equation reduces to

$$\nabla\cdot\vec{u}=0 \tag{2.1}$$

The momentum equation is obtained by applying Newton's second law to a fluid particle, *i.e.* the rate of change of momentum equals the sum of the forces acting on the particle. These are body (volume) forces and surface forces (exerted by the neighboring fluid particles) (Batchelor, 1967 ; Lamb, 1932) . Thus

$$\delta m\frac{D\vec{u}}{Dt}=[body\ forces]+[surface\ forces]$$

The body force is due to gravity and equals $\delta m\vec{g}$. The surface forces are described in terms of a stress tensor $\vec{\vec{\sigma}}$ so that the force exerted on a small surface dS oriented by a normal unit vector $\vec{n}$ is a linear function of $\vec{\vec{\sigma}}$, $\vec{n}$ and dS (Batchelor, 1967). By considering the torque exerted on an infinitesimal volume, it can be shown that the stress tensor must be symmetric. For most fluids of practical interest the expression for the stress tensor is (Batchelor, 1967)

$$\vec{\vec{\sigma}}=-p\vec{\vec{\delta}}+\mu[(\nabla\vec{u}+\nabla\vec{u}^{\mathrm{T}})-\frac{2}{3}(\nabla\cdot\vec{u})\vec{\vec{\delta}}]$$

in which p is the pressure, μ the dynamic coefficient of viscosity and $\vec{\vec{\delta}}$ the identity tensor. This is the Newton-Stokes viscosity law. Fluids satisfying this law are said to be Newtonian. Integrating the surface forces over the surface of the fluid particle and applying the divergence theorem, the momentum equation becomes

$$\frac{\partial\vec{u}}{\partial t}+\vec{u}\cdot\nabla\vec{u}=-\nabla\varphi+\nu\nabla^2\vec{u}+\vec{g} \tag{2.2}$$

In this equation, $\nu\equiv\mu/\rho$ is the kinematic coefficient of viscosity, and $\varphi\equiv p/\rho$ is pressure divided by (constant) density (*i.e.* the energy per unit mass due to the pressure field). Both are used to express the surface forces in terms of the velocity field, rather than in terms of the momentum field.

Equations (2.1) and (2.2) are called the Navier-Stokes equations for incompressible flows. The dependent variables are the velocity $\vec{u}$ and the pressure p (or φ). The theory of the Navier-Stokes equations (*abbr.* NS) is still incomplete. For instance, in the three-dimensional case, it is not known if the solution can develop singularities after a certain finite time, which shrinks to zero as the Reynolds number increases to infinity (Chorin & Marsden, 1979; Temam, 1983). The nonlinear character of the advection term $\vec{u}\cdot\nabla\vec{u}$ results in the production and maintenance of instabilities, which for high Re excite a very wide range of scales in the motion. The lack of solution regularity and the unfeasibility of defining precise initial and boundary conditions combine themselves so that the flow becomes random, *i.e.* turbulent.

2.3 Initial and boundary conditions.

To define a well-posed problem, data must be supplied so that, together with the governing partial differential equations (*abbr.* PDEs), the solution exists, is unique, and continuously dependent on the data. Such data are initial and boundary conditions.

Let Ω be a fluid domain, bounded by Σ, inside which we consider a material surface moving with the fluid whose equation is $F(\vec{x},t)=0$ (figure 2.1) . First, boundary conditions on Σ are considered. Then boundary conditions on any point $\vec{x}$ of the material surface are described. These latter are called free-surface boundary conditions when $F(\vec{x},t)=0$ represents a surface across which the properties of the medium are discontinuous (*i.e.* $F(\vec{x},t)=0$ is a free-surface).

Boundary conditions on Σ.

On Σ, one or more components of $\vec{u}$ can be prescribed (Cuvelier et al. , 1986; Gresho & Sani, 1987), *i.e.*

$$\vec{u}(\vec{x},t) = \vec{w}(\vec{x},t) \quad \vec{x}\in\Sigma \quad t\geq 0 \tag{2.3}$$

$$\int_{\Sigma} \vec{w}\cdot\vec{n}\, dS = 0 \quad t\geq 0 \tag{2.4}$$

In (2.3)–(2.4), $\vec{w}$ is the prescribed velocity on Σ. The last of these equations results from applying the divergence theorem to the continuity equation and sets an integral constraint on the normal component of the velocity on Σ. It is also possible to specify Neumann-type BCs on velocity, which are conditions for the normal and/or tangential components of the stress. These conditions take the form

$$\sigma_n = (\vec{\vec{\sigma}}\cdot\vec{n})\cdot\vec{n} \qquad \vec{x}\in\Sigma\ ; \quad t\geq 0 \tag{2.5}$$

$$\vec{\sigma}_t = (\vec{\vec{\sigma}}\cdot\vec{n})\cdot\vec{t} \qquad \vec{x}\in\Sigma\ ; \quad t\geq 0 \tag{2.6}$$

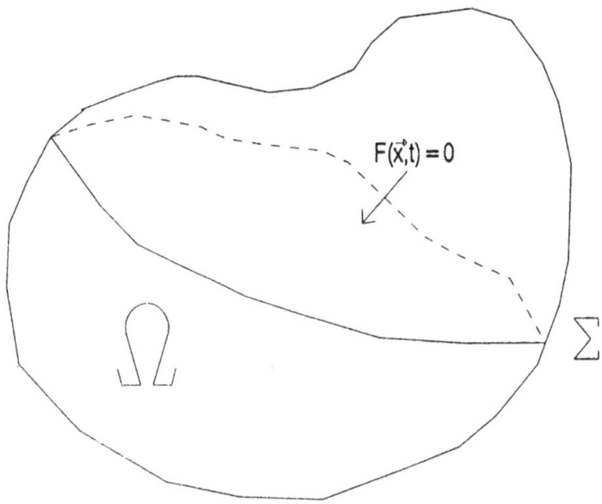

Figure 2.1 : *Solution domain for the fluid flow problem.*

If the flow is two dimensional and the boundary has a small curvature, the two stress conditions can be written as

$$\sigma_n = -p + 2\mu\frac{\partial u_n}{\partial n} \quad ; \quad u_n = \vec{u}\cdot\vec{n}$$

$$\sigma_t = \mu(\frac{\partial u_n}{\partial t} + \frac{\partial u_t}{\partial n}) \quad ; \quad u_t = \vec{u}\cdot\vec{t}$$

In three dimensions, there are two local, linearly independent, tangential directions. Provided one BC is always prescribed for each velocity component at any boundary point, both Dirichlet or Neumann conditions can be chosen, in either normal or tangential directions; it is also possible to specify different types of BCs on different parts of Σ.

An important case of a stress condition is the so-called free-slip condition, in which the tangential stress, rather than the tangential velocity, is set to zero. This boundary condition is not appropriate for real fluids, but is useful in the numerical approximations to the NS equations when the viscosity is small and the boundary layers near the walls cannot be resolved (Welch et al. , 1966; Nichols et al. , 1980).

Free-surface boundary conditions (*abbr*. FSBCs).

A free-surface is a material surface across which the density is assumed to be discontinuous.

A material surface always consists of the same particles. The rate at which any function $F(\vec{x},t)$ varies for a moving particle is given by the material derivative of F. Thus the kinematical relation defining a surface moving with the fluid is

$$\frac{DF}{Dt} \equiv \frac{\partial F}{\partial t} + \vec{u}\cdot\nabla F = 0 \tag{2.7}$$

Conversely, by requiring the motion to be continuous, it can be shown (Lamb, 1932) that a surface satisfying this equation always consists of the same particles. Fixed or moving walls also satisfy this equation, but in general (2.7) should be regarded as another equation to be solved if the position of the material surface F cannot be known *a-priori*.

The existence of a free-surface poses three problems. The first is the determination of its position. The second is the determination of its time evolution. The third is the prescription of correct BCs on all points of the free-surface. Since these boundary conditions depend on the location and shape of the free-surface, the three problems are related.

At a material surface, two transition relationships must be verified. One is the continuity of velocity. The other is the continuity of the stress vector. Continuity of velocity is a purely kinematical constraint, and is called the kinematical boundary condition. It is mathematically expressed by (2.7). Continuity of the stress vector (equal to $\vec{\vec{\sigma}}\cdot\vec{n}$) is also required to prevent the material surface from acquiring an infinite acceleration, and is called the dynamic boundary condition. The continuity of the tangential component of the stress vector is expressed by the following equality

$$\mu\left(\frac{\partial u_i}{\partial x_j}+\frac{\partial u_j}{\partial x_i}\right) n_i\, t_j \Big|_{medium\ 1} = \mu\left(\frac{\partial u_i}{\partial x_j}+\frac{\partial u_j}{\partial x_i}\right) n_i t_j \Big|_{medium\ 2} \tag{2.8}$$

In three-dimensional problems this equation must be satisfied along two different directions defined by linearly independent vectors $\vec{t}$ lying on the local tangent plane to F. In two-dimensional problems, (2.8) is a scalar equation.

The continuity of the normal component of the stress vector, gives the following equation

$$p-\mu\left(\frac{\partial u_j}{\partial x_i}+\frac{\partial u_i}{\partial x_j}\right) n_i n_j \Big|_{medium\ 1} = p-\mu\left(\frac{\partial u_j}{\partial x_i}+\frac{\partial u_i}{\partial x_j}\right) n_i n_j + p_s \Big|_{medium\ 2} \tag{2.9}$$

where p_s is the surface tension pressure. Surface tension is important in capillary waves, which sustain wind stress on the sea, and the combination of surface tension and air entrainment originates much of the complexity of aerated regions in a breaking flow. Such effects are too complex to be included in the present formulation. Therefore, for the purposes of this work, surface tension is neglected.

If the free-surface separates fluids of very different densities, the dynamic FSBCs can be simplified. This is the case in a liquid-gas interface, in which the density and viscosity of the gas are much smaller than those of the liquid. Consequently the pressure variations in the gas are much smaller if the velocities and their derivatives have comparable magnitudes. Thus, the following approximate equations may be used

$$\mu(\frac{\partial u_i}{\partial x_j}+\frac{\partial u_j}{\partial x_i})n_i t_j = 0 \tag{2.10}$$

$$p-\mu(\frac{\partial u_i}{\partial x_j}+\frac{\partial u_j}{\partial x_i})n_i n_j = p_0 \tag{2.11}$$

These equations are the tangential and normal stress conditions, respectively. The left hand side of (2.10) and (2.11) refers to conditions in the liquid, whereas the right hand side refers to conditions in the gas. If the liquid is treated as an ideal fluid, the normal stress condition is

$$p = p_0. \tag{2.12}$$

This is called the inviscid free-surface normal stress condition. It is a good approximation for the problems to be considered in this work.

Initial conditions.

Together with suitable boundary conditions (*abbr.* BCs), an appropriate initial condition (*abbr.* IC) must also be prescribed. In general this can be specified as follows

$$\vec{u}(\vec{x},t) = \vec{u}_0(\vec{x}) \qquad \vec{x} \in \bar{\Omega} \equiv \Omega \cup \Sigma\ , \quad t=0 \tag{2.13}$$

$$\nabla\cdot\vec{u}_0(\vec{x}) = 0 \qquad \vec{x} \in \Omega\ , \quad t=0 \tag{2.14}$$

$$\vec{u}_0\cdot\vec{n} = \vec{w}(\vec{x},0)\cdot\vec{n} \qquad \vec{x} \in \Sigma\ , \quad t=0 \tag{2.15}$$

Equation (2.14) expresses the constraint imposed by the continuity equation on the initial condition. Equation (2.15) is the compatibility requirement between (2.13) and (2.3). Continuity of the tangential component is not required; but if it is continuous, the solution will be smoother (Gresho & Sani, 1987).

2.4 The energy equation.

The energy density (E) of a flow field has two contributions: specific internal energy (e) due to molecular agitation, and specific kinetic (translational) energy ($\frac{1}{2}\vec{u}\cdot\vec{u}$) due to the organized fluid motion.

The equation for the rate of change of the specific internal energy results from the application of the first principle of thermodynamics to a fluid particle. If the fluid is assumed incompressible and the density uniform, the specific internal energy is not a basic dependent variable. Thus, the equation for the rate of change of the specific internal energy is not further considered.

The equation for the rate of change of $\frac{1}{2}\vec{u}\cdot\vec{u}$ is derived from the momentum equation by taking the dot product with $\vec{u}$. The result is

$$\frac{\partial}{\partial t}(\frac{1}{2}u^2)+\vec{u}\cdot\nabla(\frac{1}{2}u^2)=\vec{g}\cdot\vec{u}-\nabla\cdot(\vec{u}\varphi)+\nu\nabla^2(\frac{1}{2}u^2)-\nu\nabla\vec{u}:\nabla\vec{u} \qquad (2.16)$$

The left hand side of (2.16) is the material derivative of $\frac{1}{2}\vec{u}\cdot\vec{u}$. The first term on the right hand side represents the work done by the external force field. The second term on the right hand side expresses the work done by the pressure forces. The last two terms represent viscous diffusion and viscous dissipation of mechanical energy, respectively. Mechanical energy is conserved except for the last term. If (2.16) is integrated over a closed, periodic or infinite volume, all terms but $-\nu\nabla\vec{u}:\nabla\vec{u}$ vanish (Tennekes & Lumley, 1972). This term represents a sink of mechanical energy, which is converted into heat by the action of viscous forces. An equivalent form of the energy equation is

$$\frac{\partial}{\partial t}(\frac{1}{2}u^2)+\vec{u}\cdot\nabla(\frac{1}{2}u^2)=\vec{g}\cdot\vec{u}-\nabla\cdot(\vec{u}\varphi)+\nu\nabla\cdot(\vec{u}\times\vec{\omega})-\nu\vec{\omega}\cdot\vec{\omega} \qquad (2.17)$$

where $\vec{\omega}$ is the vorticity (§ 2.5). This form shows the important connection between vorticity and dissipation. For incompressible and homogeneous fluids, the energy balance is not essential for the description of the flow. However, it plays an important role when the mass and momentum equations are averaged in order to separate large from small-scale phenomena (§ 2.8). In that case, the energy balance usually provides an equation with independent meaning which can be used to formulate turbulence (small-scale) closure models.

2.5 The vorticity equation.

"*Why is the vorticity,* $\vec{\omega}=\nabla\times\vec{u}$ *so important ? It is not easily measurable ... it appears from its definition to have no particular physical significance; nothing seems to distinguish it especially from other gradients of the velocity field. Yet, at least for the motion of homogeneous incompressible fluids, it is the property of the flow field of crucial import, and it is not an exaggeration to say that all problems of such flows can be posed as questions about strength and location of vorticity. There are, as is well known, two principal reasons for this. Firstly, if* $\vec{\omega}=\vec{0}$ *everywhere in an incompressible fluid, then the fluid really ceases to be a fluid; it looses its infinite number of degrees of freedom, which make possible the infinite variety of fluid motion, and becomes a flexible extension of the bodies whose movement generates the flow; bring the walls to rest and the fluid stops immediately...*

...The mathematical cause of the loss of fluidic properties is the uniqueness properties of Laplace's equation; the physical reason is the relation of vorticity to the local spin, e.g. an infinitesimal cube does not rotate. Secondly, the distribution of vorticity is often compact, even though the velocity and pressure fields extend everywhere, and remains so by virtue of the Helmholtz laws of vortex motion, apart from viscous diffusion..."

P.G. Saffman, 1981

Vorticity is defined as the *curl* of velocity, $\vec{\omega} = \nabla \times \vec{u}$. It is a fundamental component of the local velocity gradient, describing the rate of rotation of fluid particles. Specifically, $\vec{\omega}$ is twice the local angular velocity. Because of this, it can be used to determine if a flow field is "really a flow field" or a "flexible solid".

One reason why vorticity is important, is that it can be used to distinguish laminar from turbulent flows. In laminar flows, vorticity is usually concentrated in thin regions near the boundaries (Batchelor, 1967; Saffman, 1981). Turbulent flows, on the other hand, are characterized by large amounts of random, fluctuating vorticity distributed in a wide range of space and time scales (Tennekes & Lumley, 1972). Another reason why vorticity is important, is that under certain circumstances it is advantageous to use it as a dependent variable in the numerical approximations of the incompressible flow equations.

The equation describing the rate of change of vorticity for an incompressible, constant-property fluid is derived by taking the *curl* of the momentum equation. Making use of the identity $\nabla \times (\vec{u} \cdot \nabla \vec{u}) = \vec{\omega}(\nabla \cdot \vec{u}) - \vec{u}(\nabla \cdot \vec{\omega}) + \vec{u} \cdot \nabla \vec{\omega} - \vec{\omega} \cdot \nabla \vec{u}$ and of the solenoidal property of $\vec{u}$ and $\vec{\omega}$, and assuming body forces to be conservative, it is found that

$$\frac{\partial \vec{\omega}}{\partial t} + \vec{u} \cdot \nabla \vec{\omega} = \vec{\omega} \cdot \nabla \vec{u} + \nu \nabla^2 \vec{\omega} \qquad (2.18)$$

The left hand side of this equation is the material derivative of $\vec{\omega}$. Likewise, the term $\nu \nabla^2 \vec{\omega}$ is the rate of change of $\vec{\omega}$ due to viscous diffusion. The term $\vec{\omega} \cdot \nabla \vec{u}$ is known as the vortex-stretching term (Batchelor, 1967; Tennekes & Lumley, 1972). It does not have a counterpart in the momentum or energy equations, and shows that the vorticity of a fluid element may be amplified if the fluid element is stretched (by the velocity gradient) in the direction of $\vec{\omega}$. This is a consequence of the principle of conservation of angular momentum. The vortex-stretching term is fundamental for explaining the dynamics of turbulence. It is the stretching mechanism that transfers energy from the large structures in the flow, to the smallest structures, where it is dissipated by the action of viscosity. Pressure is absent in the vorticity balance. This is because $\nabla \times \nabla \varphi$ vanishes identically; in other words, because angular momentum cannot change by the action of normal forces.

The fluid flow problem is considerably simplified if it is assumed that $\vec{\omega}=\vec{0}$ everywhere. In that case, if the fluid domain is simply connected, a potential function ϕ exists such that $\vec{u}=\nabla\phi$. Substituting into the continuity equation gives $\nabla^2\phi=0$. This equation (Laplace) is linear, and the constraints on the possible solutions are so strong that the fluid looses its "fluidic properties". The spatial distribution of $\vec{u}$ is forced to satisfy simple linear equations, independent of temporal changes. Using Gauss's theorem, it can be concluded that for a simply connected fluid domain

$$\int_{\Omega} \vec{u}\cdot\vec{u}\, dV = \int_{\Sigma} \phi\vec{u}\cdot\vec{n}\, dS$$

and if at some instant the normal component of velocity is zero at all boundary points, $\vec{u}=\vec{0}$ everywhere. If the flow is incompressible and irrotational, energy conservation is expressed by Bernoulli's equation

$$\frac{\partial\phi}{\partial t}+\frac{1}{2}(u^2+v^2)+\varphi-\vec{g}\cdot\vec{x}=R(t) \tag{2.19}$$

Potential (irrotational) flow is the basis for most wave theories used in hydraulics and coastal engineering. This is both the strength and the weakness of such theories. Strength, because the mathematical treatment is greatly simplified. Weakness, because the flow in breaking waves and hydraulic jumps, is rotational. This question occupies much of chapter 3.

2.6 The pressure Poisson equation for incompressible flows.

In an incompressible flow the pressure φ is determined through the equation $\nabla\cdot\vec{u}=0$, which in the numerical approximations is usually replaced by a pressure Poisson equation (*abbr.* PPE).

Any vector field $\vec{v}$ in a region Ω with smooth boundary Σ can be uniquely decomposed in the form (Chorin & Marsden, 1979)

$$\vec{v}=\vec{v}_o+\nabla\phi_p$$

where $\vec{v}_o$ has zero divergence and is parallel to the boundary, *i.e.* $\vec{v}_o\cdot\vec{n}=0\ \forall\vec{x}\in\Sigma$. Let $\mathbb{P}$ and $\mathbb{Q}$ be two operators which project $\vec{v}$ onto the nullspace of the *div* and *curl* operators, respectively (*i.e.* $\nabla\cdot\mathbf{P}=0$ and $\nabla\times\mathbb{Q}=0$ symbolically). Writing the momentum equation in the form $\partial\vec{u}/\partial t+\nabla\varphi=\vec{f}(\vec{u})$; $\vec{f}(\vec{u})=\nu\nabla^2\vec{u}-\vec{u}\cdot\nabla\vec{u}+\vec{g}$, it is found that

$$\frac{\partial\vec{u}}{\partial t}=\mathbb{P}(\vec{f}(\vec{u})) \quad ; \quad \nabla\varphi=\mathbb{G}(\vec{f}(\vec{u}))$$

because $\nabla\cdot(\partial\vec{u}/\partial t)=\partial/\partial t(\nabla\cdot\vec{u})=0$ and $\nabla\times\nabla\varphi=0$. The fact that the local acceleration

$\partial\vec{u}/\partial t$ is divergence-free implies that the pressure can be defined by the solution of the following Poisson-Neumann problem

$$\nabla^2\varphi = \nabla\cdot\vec{f} \tag{2.20}$$

$$\vec{n}\cdot\nabla\varphi = \vec{n}\cdot\vec{f} \tag{2.21}$$

An extensive review of the possible forms for the PPE (2.20) and associated pressure boundary condition (2.21) can be found in Gresho & Sani (1987). The PPE is the basis for a number of algorithms which step alternately between velocity and pressure. Some of these algorithms are discussed in chapter 4. The combination momentum equation+PPE is not completely equivalent to the original NS problem. The PPE requires a greater degree of regularity and smoothness on $\nabla\varphi$ and $\vec{u}$ than the NS system.

Remarks

1. The pressure is automatically eliminated in the momentum equation if the form of the projection operator $\mathbb{P}$ is known.

2. The explicit formal expansion of $\mathbb{P}$ and $\mathbb{G}$ is (Gresho, 1990)

$$\mathbb{P} = [I - \nabla(\nabla^2)^{-1}\nabla\cdot]$$

$$\mathbb{G} = [\nabla(\nabla^2)^{-1}\nabla\cdot]$$

where I is the identity operator. These operators are difficult to use directly, because ∇^2 is formally inverted by a Green function, which must have "built-in" information about the particular BC's and domain geometry (and therefore cannot be regarded as a trivial entity).

2.7 General aspects of turbulent flows. Averaging methods and Reynolds equations.

When a free-surface flow breaks, zones with a strong turbulence level are created. The turbulence level is maximum close to the free-surface, where internal recirculation and air friction act as sources, and decays downstream and downwards of the breaking front, where dissipation outweighs production.

One problem towards realistic modeling of the flow structure in this type of flow, is the simultaneous coexistence of zones with a very high turbulence level and zones where the flow is almost irrotational. At present, there is no uniformly valid theoretical way of treating such abrupt transitions; for the schemes used for the representation of strong turbulence are not adequate for low-intensity turbulence and transition to a laminar regime, and vice versa.

It is difficult to give a definition of turbulence. However, it is possible to indicate some characteristics that are common to all turbulent flows :

- Irregularity : The dependent variables are random functions of space and time. Their averaged values can be calculated, but the random fluctuations must be treated statistically.

- High diffusivity : Mixing of properties in a turbulent flow is much greater than mixing due to molecular diffusion. If $\mathcal{U}$ and $\mathcal{L}$ are typical scales of velocity and length for the velocity perturbations, this implies that :

$$Re = \frac{\mathcal{U}\mathcal{L}}{\nu} \sim \frac{Turbulent\ diffusivity}{Molecular\ diffusivity} \gg 1$$

- Three-dimensional vorticity fluctuations : Turbulence is originated in the nonlinear vortex stretching mechanism, which transfers energy from large scales to small scales.

- Dissipation : Turbulence extracts energy from the mean flow and needs a source to be maintained. This source can be a velocity shear, buoyancy, magneto-hydrodynamic effects, etc.

- Turbulence is flow-dependent, not fluid-dependent : Turbulence diffusivity varies from point to point with time, and is influenced by external conditions (configuration of the fluid domain, roughness of confining walls, etc); it is not described in terms of a fluid property.

The NS equations are valid for both laminar and turbulent flows, but the statistical approach to turbulence is the only practical way of treating the latter. This may be seen by using the Kolmogorov scales for the length, time and velocity

$$l_o \equiv (\nu^3/\epsilon)^{1/4} \quad ; \quad t_o \equiv (\nu/\epsilon)^{1/2} \quad ; \quad v_o \equiv (\nu\epsilon)^{1/4}$$

to estimate the number of degrees of freedom, N, required to describe a turbulent flow in a domain with typical linear dimension l_d in n dimensions :

$$N = \left(\frac{l_d}{l_o}\right)^n \sim \left(\frac{\mathcal{U}\mathcal{L}}{\nu}\right)^{\frac{3n}{4}} = (Re)^{\frac{3n}{4}}$$

With $Re = 10^5$, this gives $N \sim 10^9$, which is beyond the capacities of the most powerful computers presently available. Therefore, statistical averaging is applied to the NS equations, and predictions are based on averaged properties, which vary much more gradually in space and time.

For the purposes of the present work all averages are assumed to be ensemble averages, because ensemble averaging is applicable to both steady and unsteady flows. Ensemble averaging for a generic property $\mathcal{F}(\vec{x},t)$ is defined by

$$\overline{\mathcal{F}(\vec{x},t)} = \int \mathcal{F}(\vec{x},t)\,\mathcal{D}(\vec{x},t)\,d\mathcal{V} = \lim_{N\rightarrow\infty} \frac{1}{N}\sum_{n=1}^{N} \mathcal{F}_n(\vec{x},t)$$

where $\mathcal{D}(\vec{x},t)$ is the probability density function of $\mathcal{F}(\vec{x},t)$ and the integration is taken over function space (over all possible realizations). Ensemble averages will be denoted by an overbar. The ensemble average of $\mathcal{F}$ will be denoted by F, and the instantaneous perturbation, $\mathcal{F}-F$, by f. The following rules are assumed

- $\overline{f} = 0$
- $\overline{Ff} = 0$
- $\overline{\overline{F}} = \overline{F} = F$
- partial derivatives can be commuted with the averaging operator.

Equations for the time evolution of mean flow variables can be derived from the original NS system by replacing u_i, φ by $u_i + U_{i,\varphi} + \varphi$ and averaging[1]. This gives the Reynolds equations

$$\frac{\partial U_i}{\partial x_i} = 0 \tag{2.22}$$

$$\frac{\partial U_i}{\partial t} + U_j \frac{\partial U_i}{\partial x_j} = -\frac{\partial \varphi}{\partial x_i} + \nu \frac{\partial^2 U_i}{\partial x_j^2} + g_i - \frac{\partial \overline{u_i u_j}}{\partial x_j} \tag{2.23}$$

These equations are identical to the primitive NS system except for the last term in (2.23), which expresses the influence of the small scales in the mean flow. The tensor $\sigma_{ij}^r = -\rho\overline{u_i u_j}$ is the Reynolds stress tensor. The system (2.22)–(2.23) contains more unknowns than equations, and thus is not well-posed. This is known as the closure problem. Additional empirical assumptions must be supplied to express the velocity correlations $\overline{u_i u_j}$ in terms of

[1] *In this work, instantaneous values of the dependent variables in the (NS and related) equations are denoted by lower-case letters. In turbulent flows, the averaged quantities are denoted by upper-case letters, and the fluctuations by lower-case letters. The meaning of symbols should always be clear from the context.*

knowable quantities. The set of equations that performs this task is called a turbulence model.

The most widely used closure scheme is Boussinesq's eddy-viscosity (ν_t) concept, which assumes that the Reynolds stresses are proportional to the rate of deformation of the mean flow

$$\overline{u_i u_j} = -\nu_t\left(\frac{\partial U_i}{\partial x_j}+\frac{\partial U_j}{\partial x_i}\right)+\frac{2}{3}K\delta_{ij} \tag{2.24}$$

In this equation, ν_t is the eddy-viscosity coefficient and $K=\frac{1}{2}\overline{u_i u_i}$ is the turbulent kinetic energy (per unit mass). The last term in this equation ensures that the sum of the normal turbulent stresses has the correct value, and is usually included in φ. ν_t is not a fluid property, and may vary in space and time, depending on the local state of turbulence.

Using (2.24), the task of the turbulence model consists of determining a distribution of ν_t so that the mean flow is correctly described. For a scalar quantity $\mathcal{F}$, the equivalent of (2.24) is

$$\overline{u_i f} = -\Gamma_t\frac{\partial F}{\partial x_i}$$

and Γ_t is called the eddy-diffusivity.

The eddy-viscosity concept is not the only possible solution for the closure problem. Other alternatives consist of deriving transport equations for the Reynolds stresses, or applying a spatial filter to the NS equations. These alternatives are more complex and expensive than (2.24) for engineering purposes, and are not free from empirical assumptions. Consequently, this work is focussed on the eddy-viscosity approach.

In complicated turbulent flows, information on the local state of turbulence must be supplied in order to describe the distribution of ν_t, *i.e.* the turbulent transport processes must be taken into account (see figure 2.2). These processes may be interpreted by analyzing the exact transport equations for some relevant turbulence-related quantities (§ 2.8). Using the information provided by such equations, a turbulence model that is consistent with the physical requirements of our problem is summarized (§ 2.9).

2.8 Turbulence transport equations.

One way to describe the local state of turbulence, consists of deriving and analyzing transport equations for turbulence-related quantities. Such equations are obtained from the NS equations. The exact transport equations cannot be used without simplifications, as they also contain unknown correlations; however, the physical mechanisms affecting the quantity under consideration can be identified by inspecting the form of the various terms they contain.

The turbulence-related quantities in which we are interested are: the turbulent kinetic energy (per unit mass) $K=\frac{1}{2}\overline{u_i u_i}$; the rate of dissipation of K, defined by $\epsilon=\nu\overline{(\partial u_i/\partial x_j)^2}$; and the velocity correlations $\overline{u_i u_j}$ (which define the Reynolds stress tensor).

The turbulent kinetic energy is the most relevant turbulence velocity scale. Because most of the kinetic energy of turbulence is contained in large eddies, $K^{1/2}$ is a velocity scale for the large scale turbulent motion.

The rate of dissipation also is an important variable. It is the only term in the transport equation for K which acts as a sink, and is related to the smallest structures in the flow. These are very effective in dissipating energy and contain most of the random vorticity characteristic of turbulence. Furthermore, the equation for the rate of change of ϵ plays an important role in turbulence modeling (both in eddy-viscosity and Reynolds stress formulations).

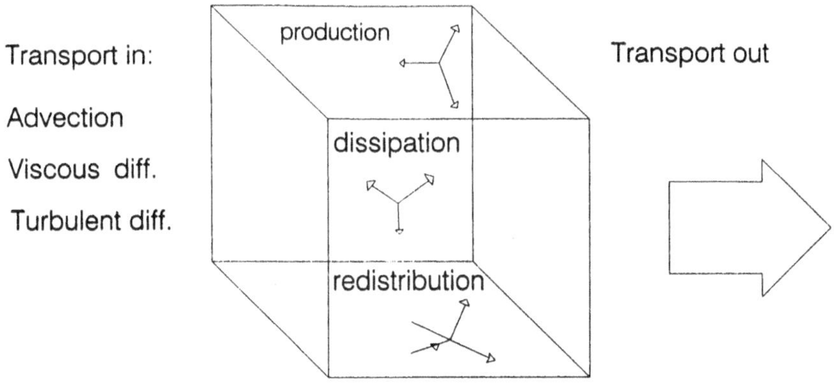

Figure 2.2 : *Turbulence transport processes.*

The transport equation for the velocity correlations $\overline{u_i u_j}$ is analyzed in order to understand the limitations of the eddy-viscosity concept; it is inherently more complex than the previously referred equations, because $\overline{u_i u_j}$ is a tensor quantity.

2.8.1 The exact K equation.

The exact equation for the turbulent kinetic energy K can be obtained by multiplying the momentum equation for the total velocity u_i+U_i by the velocity perturbation u_i , and averaging all terms (Launder & Spalding, 1972; Tennekes & Lumley, 1972). The result is

$$\frac{\partial K}{\partial t}+U_j\frac{\partial K}{\partial x_j}=-\frac{\partial}{\partial x_j}(\overline{u_j\varphi}+\overline{\tfrac{1}{2}u_i^2u_j})+\nu\frac{\partial^2 K}{\partial x_j^2}-\overline{u_iu_j}\frac{\partial U_i}{\partial x_j}-\nu\overline{(\frac{\partial u_i}{\partial x_j})^2} \tag{2.25}$$

The left hand side of (2.25) is the material derivative of K . The first two terms on the right hand side (within brackets) represent turbulent diffusion of K, due to pressure-velocity correlations and velocity-kinetic energy correlations. These terms are divergences of energy fluxes, and redistribute energy from one point to another. The second term on the right hand side of (2.25) represents transport of K by viscous stresses, *i.e.* laminar diffusion. Usually, it is much smaller than the turbulent diffusion term. The third term on the right hand side of (2.25) represents the rate at which mean kinetic energy is converted in turbulent kinetic energy, by the working of the Reynolds stresses against the mean flow gradient. In contrast with the diffusion terms, (which represent spatial redistribution) it does not necessarily vanish when integrated over a closed control volume. Thus, this term exchanges energy between the mean flow and turbulence, and normally involves a loss to the mean flow and a gain to turbulence. For that reason, it is called the production term, and will be denoted by $\mathcal{P}$. The last term of (2.25) is the rate at which viscous stresses dissipate turbulent kinetic energy, and will be denoted by ϵ. It always represents a loss of energy, and is fundamental to the dynamics of turbulence.

The K equation provides information on turbulent transport processes that is not contained in the Reynolds equations. However, this information is not usable because new unknown correlations ($\overline{u_j\varphi}$, $\overline{\tfrac{1}{2}u_i^2u_j}$ and ϵ) have been introduced.

2.8.2 The exact ϵ equation.

The exact ϵ equation can be derived by differentiating the x_i-component of the momentum equation for the total flow with respect to x_j , multiplying by the fluctuating velocity gradient $\partial u_i/\partial x_j$, contracting indices and averaging. The result is

$$\begin{aligned}\frac{\partial \epsilon}{\partial t}+U_j\frac{\partial \epsilon}{\partial x_j}=&-2\nu\overline{\frac{\partial u_i\partial u_i\partial u_k}{\partial x_k\partial x_l\partial x_l}}-2\overline{(\nu\frac{\partial^2 u_i}{\partial x_k\partial x_l})^2}-\\&-\frac{\partial}{\partial x_k}\{\overline{\nu u_k(\frac{\partial u_i}{\partial x_j})^2}+2\nu\overline{\frac{\partial u_k}{\partial x_l}\frac{\partial \varphi}{\partial x_l}}-\nu\frac{\partial \epsilon}{\partial x_k}\}-\\&-2\nu\{\overline{\frac{\partial u_i\partial u_k}{\partial x_l\partial x_l}}+\overline{\frac{\partial u_l\partial u_l}{\partial x_i\partial x_k}}\}\frac{\partial U_i}{\partial x_k}-2\nu\overline{u_k\frac{\partial u_i}{\partial x_l}}\frac{\partial^2 U_i}{\partial x_l\partial x_k\partial x_l}\end{aligned} \tag{2.26}$$

This equation contains many unknown correlations. Despite the complexity of the terms in (2.26), it is possible to interpret their physical significance, and even to construct realistic, yet crude, closure assumptions to render it tractable.

The left hand side of (2.26) is the material derivative of the quantity being transported. The first term on the right hand side of (2.26) represents production by vortex stretching due to the fluctuating velocity gradient (or, as shown by Tennekes & Lumley

(1972), the fluctuating vorticity). This is the main production term for ϵ, because it involves correlations of quantities with very high values (the exact correlation coefficient is unknown). The second term on the right hand side of (2.26) represents viscous dissipation due to the spatial gradients of the fluctuating velocity gradient (or spatial gradients of fluctuating vorticity). This term is negative-definite, does not vanish upon integration over a closed volume, and is the only one which prevents ϵ from increasing indefinitely. The terms grouped as a spatial divergence on the right hand side of (2.26) express viscous and turbulent diffusion of ϵ. Usually, the viscous diffusion terms are much smaller than their turbulent counterparts, except near walls and in stagnant regions where the turbulence intensity is small. The last two terms of (2.26) represent production due to the interaction between the turbulent correlations and the mean flow velocity gradient. They express mixed production effects.

Very little is known about the statistical behavior of the correlations in the ϵ equation, which are difficult or (at present) impossible to measure. There is experimental evidence that high values of ϵ are intermittent and concentrated in thin regions inside the flow (Tennekes & Lumley, 1972). Such regions result from the final collapse of vortex tubes stretched by the mean and fluctuating velocity fields.

In practical turbulence modeling, the ϵ equation is used both as a complement to the K equation, and as a basis upon which more complex formulations (Reynolds stress transport equations) are constructed (Hanjalić & Launder, 1976; Launder, 1979; Launder et al. , 1975).

2.8.3 The exact $\overline{u_i u_j}$ equation.

The exact equation for the $\overline{u_i u_j}$ second-order velocity correlation tensor can be derived by first taking the x_i-component of the NS equations for the total flow, multiplying by the velocity fluctuation u_j and averaging; reversing the roles of i and j; and adding the results. The result is (Bradshaw, 1978; Launder, 1979)

$$\frac{\partial}{\partial t}\overline{u_i u_j}+U_l\frac{\partial}{\partial x_l}\overline{u_i u_j}=P_{ij}+\Phi_{ij}-\frac{\partial J_{ijl}}{\partial x_l}-D_{ij} \tag{2.27}$$

where

$$P_{ij}=-\left(\overline{u_j u_k}\frac{\partial U_i}{\partial x_k}+\overline{u_i u_k}\frac{\partial U_j}{\partial x_k}\right)$$

$$\Phi_{ij}=\overline{\varphi\left(\frac{\partial u_i}{\partial x_j}+\frac{\partial u_j}{\partial x_i}\right)}$$

$$J_{ijl}=\overline{u_i u_j u_l}-\nu\frac{\partial}{\partial x_l}\overline{u_i u_j}+\overline{\varphi(\delta_{jl}u_i+\delta_{il}u_j)}$$

$$D_{ij}=-2\nu\overline{\frac{\partial u_i}{\partial x_k}\frac{\partial u_j}{\partial x_k}}$$

Different values of i and j give the six independent second-order velocity correlations, which in turn define the Reynolds stress tensor. The left hand side of (2.27) is the material derivative of $\overline{u_i u_j}$. This term is identical for any transportable property. P_{ij} is called the production tensor, and represents generation of turbulent velocity correlations. Unlike the production terms in the K and ϵ equations, which are almost always positive, "production" of $\overline{u_i u_j}$ can be of either sign (Launder, 1979). Φ_{ij} is called the pressure-strain redistribution tensor. This term tends to make turbulence more nearly isotropic, by redistributing energy among the components of the second-order velocity correlation tensor. J_{ijl} is called the transport tensor. The divergence of J_{ijl} accounts for turbulent and viscous diffusion (*i.e.* spatial redistribution) of $\overline{u_i u_j}$. D_{ij} is called the dissipation tensor. This term accounts for dissipation of $\overline{u_i u_j}$ in (2.27). If turbulence is assumed nearly isotropic, D_{ij} can be considered as a function of ϵ alone (Launder, 1979; Launder et al. , 1975).

Equation (2.27) is included in the present review with the main purpose of illustrating the limitations of the eddy-viscosity concept. Clearly, this latter is incapable of representing the complexity of the turbulent transport processes that affect the Reynolds stresses.

2.9 Turbulence models.

2.9.1 Definition and classification.

A turbulence model is a set of equations which expresses the (unknown) Reynolds stress tensor in terms of mean flow variables (mean velocity and mean pressure), and possibly in terms of other determinable quantities (related to the local state of turbulence).

The transport equations for turbulence-related quantities provide information on the physical aspects of turbulent transport phenomena that is not contained in the Reynolds equations; but the price is the introduction of many new unknown correlations. Thus, to be used as part of any turbulence model, those equations also require empirical assumptions.

Figure 2.3 summarizes the most important types of turbulence models that have been used for predicting non-homogeneous and unsteady turbulent flows.

Using the Reynolds stress philosophy, which starts with taking time or ensemble averages of the NS equations, the two main classes of models are eddy-viscosity (ν_t) models, and Reynolds stress models (also known as second-order closure models).

In eddy-viscosity models, the Reynolds stress tensor is expressed in terms of the mean velocity gradient by (2.24) . This closure equation is based on an analogy between laminar and turbulent diffusion, which in many flows is not correct. The generic expression for ν_t is

$$\nu_t = \mathcal{U}\mathcal{L}$$

where $\mathcal{U}$ is a typical scale for the velocity fluctuations and $\mathcal{L}$ is a "mixing length" (analogous to the mean free path in kinetic gas theory).

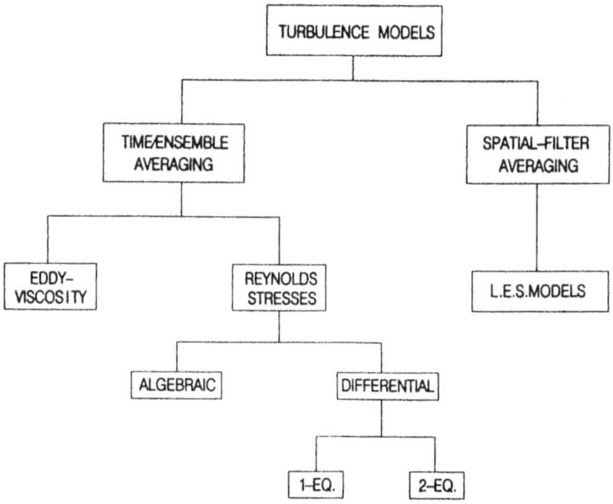

Figure 2.3 : *Classification of turbulence models.*

If $\mathcal{U}$ and $\mathcal{L}$ are determined by means of algebraic equations, the turbulence model is said to be algebraic. Algebraic models have the important advantages of economy and simplicity. However, they have the disadvantage of lacking universality. If $\mathcal{U}$ or $\mathcal{L}$ (or both) are determined by means of differential equation(s), the model is said to be differential. According to the number of equations employed, differential models can be further classified as 1-equation models, or 2-equation models. Differential eddy-viscosity models have the advantage of greater universality over algebraic models; however, they have the disadvantage of being more computationally expensive.

It is also possible to abandon the eddy-viscosity concept, and formulate closed-form equations for the Reynolds stress tensor. Turbulence models based on this approach are called Reynolds stress closures (or second-order closures). Such models are sometimes called "multi-equation models", due to the fact that they are usually composed of more than two equations. Reynolds-stress turbulence models have the advantage of describing turbulent processes with a degree of generality not possessed by eddy-viscosity models; this is the reason why they give better predictions in complicated flows (Launder, 1989). However, such models also have disadvantages: they are computationally expensive, require many (involved) closure assumptions, are more difficult to calibrate, and rely in part on simpler equations (closed-form K and ϵ equations).

Another distinct approach to turbulence consists of applying a spatial filter to the NS equations, instead of averaging over time (or realizations). In this way, all perturbations with a spatial scale smaller than the width of the filter are eliminated. This approach is known as

Large-eddy simulation (L.E.S.) (Leonard, 1974). The greatest advantage with respect to conventional turbulence models, is that such type of averaging assures a better preservation of turbulence properties. However, closure assumptions are also required in order to describe the influence of the small scales on the large scales; these are known as subgrid-scale closures. Though promising, this type of turbulence modeling is computationally expensive, and has not been used on a routine basis in engineering applications.

A brief comparative review of turbulence models is now presented, to clarify the merits and demerits of each one, and justify the choice of the K-ϵ model as the basis for representing turbulence in recirculating free-surface flows.

2.9.2 Algebraic models.

For the purposes of the present work, the only relevant algebraic model is the mixing-length model; it is used near solid walls in replacement of the more complex differential model. If dU/dy is the only relevant velocity gradient near a wall, the mixing length model assumes the velocity scale $\mathcal{U}$ to be proportional to the velocity shear and to a "mixing length" l_m . Thus

$$\nu_t = l_m^2 \left| \frac{dU}{dy} \right|$$

The choice of the mixing length depends on the characteristics of the flow. In the case of interest, it may be chosen proportional to the distance away from the wall

$$l_m = \kappa y$$

where $\kappa \simeq 0.41$ is the von-Kármán constant. Very near the wall, where viscous effects must be taken into account, the mixing length is best prescribed using van-Driest's formula (Launder & Spalding, 1972)

$$l_m = \kappa y [1 - exp(-\frac{y u_*}{A \nu})]$$

where u_* is the friction velocity, $(|\tau_w|/\rho)^{1/2}$, and A is van-Driest's constant , $A \simeq 26$. This formula approaches the pure mixing length distribution away from the wall; close to the wall, the mixing length is reduced to reflect viscous damping.

It is possible to show (*e.g.* Rodi, 1980) that the mixing length model implies that turbulence is assumed to be in a state of local equilibrium between production and dissipation, and convection and diffusion of turbulence properties is neglected. Thus, it is inadequate for our problem, and a more sophisticated approach is required. Nevertheless, the mixing length model is implicitly used near solid walls, where the differential equations governing the evolution of turbulence must be replaced by algebraic relationships between the tangential mean flow velocity and the turbulence-related variables (§ 2.10.1) .

2.9.3 The K (one-equation) model.

The turbulent kinetic energy K is a direct measure of the intensity of the velocity fluctuations, *i.e.* $\mathcal{U} \simeq \sqrt{K}$. The space distribution and time evolution of K is described by (2.25). However, (2.25) cannot be used because of the unknown correlations in the turbulent diffusion, production and dissipation terms. To obtain a closed system of equations, model assumptions must be introduced.

The production term is modeled using (2.24) . The turbulent diffusion term is modeled using a gradient-diffusion hypothesis (Launder & Spalding, 1972; Rodi, 1980). The rate of dissipation term is modeled assuming local isotropy of the small scale structures (which are responsible for most of the dissipation) and dimensional analysis. Using these closure assumptions, the modeled K equation can be written as (Rodi, 1980)

$$\frac{\partial K}{\partial t}+U_j\frac{\partial K}{\partial x_j}=\frac{\partial}{\partial x_j}\left(\frac{\nu_t}{\sigma_k}\frac{\partial K}{\partial x_j}\right)+\nu_t\left(\frac{\partial U_i}{\partial x_j}+\frac{\partial U_j}{\partial x_i}\right)\frac{\partial U_i}{\partial x_j}-C_d\frac{K^{3/2}}{l_t} \tag{2.28}$$

In this equation, σ_k is an empirical diffusion constant (turbulent Prandtl number for K, which is of $\mathcal{O}(1)$), and C_d a further empirical constant. The dependent variable is K; l_t is a turbulent length scale that must be prescribed algebraically. In (2.28), it was assumed that molecular diffusion of K is much smaller than turbulent diffusion; otherwise, if viscous effects are significant (*e.g.* near walls, in zones of low turbulence intensity, etc), molecular diffusion should be included, and the last term in (2.28) modified to account for direct viscous dissipation.

The K model has the advantage of being more universal than algebraic models, because it takes into account turbulent transport processes in the definition of the velocity length scale $\mathcal{U}$. However, it also presents the disadvantage of requiring algebraic prescriptions for the length scale l_t ; in complex flows (*e.g.* recirculating flows) this is impossible, because the length scale is also affected by the same transport effects as K itself. Thus, to predict such flows successfully, the K equation needs to be complemented by at least another equation which removes the need to prescribe l_t by means of algebraic formulae.

2.9.4 The two-equation K-ϵ model.

In § 2.9.3, it was seen that the K equation needs to be supplemented by another equation which takes into account the transport processes affecting l_t. In practice, it is advantageous to use ϵ as second independent variable, in place of l_t (Rodi, 1980). Assuming isotropic dissipation (since the small eddies that do most of the dissipation do not have preferential orientations, if the Reynolds number is high) it is found that

$$\nu_t = C_d \frac{K^2}{\epsilon} \tag{2.29}$$

by dimensional analysis. $C_d \simeq 0.09$ is an empirical constant; $\frac{K}{\epsilon}$ is a "characteristic decay time" (say, $\mathcal{T}$), and thus $\nu_t \sim \mathcal{U}^2\mathcal{T} = \mathcal{U}\mathcal{L}$.

The exact ϵ equation requires closure hypotheses for the diffusion, production and dissipation terms. The diffusion term is modeled using a gradient diffusion assumption

$$Diffusion\ of\ \epsilon \rightarrow \frac{\nu_t}{\sigma_\epsilon}\frac{\partial \epsilon}{\partial x_j}$$

where σ_ϵ is an empirical diffusion constant. The production and dissipation terms are not modeled separately; only their difference is modeled (Rodi, 1980)

$$(Production - Dissipation)_\epsilon \rightarrow C_{1\epsilon}\frac{\epsilon}{K}\mathcal{P} - C_{2\epsilon}\frac{\epsilon^2}{K}$$

$$\rightarrow \frac{1}{(Decay\text{-}time)} \times (C_{1\epsilon}\mathcal{P} - C_{2\epsilon}\epsilon)$$

where $\mathcal{P}$ is the production of K. The final modeled ϵ equation, proposed by Launder & Spalding (1972), Rodi (1980) and others is

$$\frac{\partial \epsilon}{\partial t} + U_j \frac{\partial \epsilon}{\partial x_j} = \frac{\partial}{\partial x_j}\left(\frac{\nu_t}{\sigma_\epsilon}\frac{\partial \epsilon}{\partial x_j}\right) + C_{1\epsilon}\frac{\epsilon}{K}\nu_t\left(\frac{\partial U_i}{\partial x_j} + \frac{\partial U_j}{\partial x_i}\right)\frac{\partial U_i}{\partial x_j} - C_{2\epsilon}\frac{\epsilon^2}{K} \tag{2.30}$$

If ϵ is computed with (2.30), the K equation becomes

$$\frac{\partial K}{\partial t} + U_j \frac{\partial K}{\partial x_j} = \frac{\partial}{\partial x_j}\left(\frac{\nu_t}{\sigma_k}\frac{\partial K}{\partial x_j}\right) + \nu_t\left(\frac{\partial U_i}{\partial x_j} + \frac{\partial U_j}{\partial x_i}\right)\frac{\partial U_i}{\partial x_j} - \epsilon \tag{2.31}$$

The dependent variables in (2.31) and (2.30) are K and ϵ , which define ν_t by (2.29). Equations (2.31) and (2.30) constitute the two-equation K-ϵ model. σ_ϵ is the Prandtl/Schmidt number for ϵ, whose recommended value is 1.3 (Rodi, 1980). The K-ϵ model requires six empirical constants, whose recommended values are presented in table 2.1 .

The main advantages of the K-ϵ model are :

- ν_t is completely defined in terms of transport equations.

- The model is successful in predicting some complex flows, such as recirculating flows.

- The model is more universal than simpler models, and the task of calibrating the empirical constants is relatively moderate.

- It can be improved (*i.e.* including anisotropic turbulent diffusion, low-intensity turbulence, streamline curvature effects, etc).

Table 2.1 : Recommended values for the empirical constants of the K-ϵ turbulence model

C_d	σ_k	σ_ϵ	C_1	C_2
0.09	1.0	1.3	1.44	1.92

The main disadvantages are the use of the ν_t concept (which is of doubtful value in complicated flows), and the very crude closure assumptions present in the ϵ equation; these do not work in low Reynolds number turbulent flows. From the numerical viewpoint, the ϵ equation poses difficult problems, since both production and dissipation terms are singular when $K \rightarrow 0$.

The closure assumptions of the K-ϵ model are valid only for high-intensity and nearly isotropic turbulence. In turbulent free-surface flows, there are zones with strong turbulence and zones where the flow is nearly laminar. The transition between these regions is often abrupt and intermittent. Therefore, it is necessary to refine the standard K-ϵ model to describe such transitions properly. This problem has been analyzed by several authors (*e.g.* Jones & Launder, 1972; Harlow & Nakayama, 1967,1968). In this work, the proposal of Harlow & Nakayama is followed, and the eddy-viscosity ν_t is replaced by

$$(\nu_t{}^2)\frac{[1 - exp(-\beta\frac{\nu}{\nu_t})]}{\beta\nu}$$

i.e. the effective eddy-viscosity is reduced for low-intensity turbulence. The form of the damping function is obtained by relating the mean viscous retarding acceleration of a turbulence eddy to the Oseen approximation for the force on a sphere (Hinze, 1975; Harlow & Nakayama, 1967, 1968). The cut-off is determined by the local Reynolds number ν_t/ν and by an empirical constant, whose recommended value is 100.

2.9.5 Reynolds-stress transport models.

Reynolds-stress models, based on modeled forms of the $\overline{u_i u_j}$ transport equation, are significantly more complex than eddy-viscosity models. The exact $\overline{u_i u_j}$ equation requires closure assumptions for the dissipation, diffusion, and pressure-strain (redistribution) terms. The form of those assumptions is still a matter of debate (Launder, 1979; Launder, 1989). One set of such model closures is the following

$$J_{ijl} = C_s \frac{K}{\epsilon} \overline{u_k u_l} \frac{\partial \overline{u_i u_j}}{\partial x_k}$$

$$\epsilon_{ij} = \frac{2}{3} \epsilon \delta_{ij}$$

$$\Phi_{ij} = \Phi_{ij,1} + \Phi_{ji,1} + \Phi_{ij,2} + \Phi_{ji,2}$$

$$\Phi_{ij,1} + \Phi_{ji,1} = -C_1 \frac{\epsilon}{K} (\overline{u_i u_j} - \frac{2}{3} \delta_{ij} K)$$

$$\Phi_{ij,2} + \Phi_{ji,2} = -\gamma (P_{ij} - \frac{2}{3} \delta_{ij} \mathcal{P})$$

where $\mathcal{P}$ is the production of kinetic energy (which is computed without invoking the eddy-viscosity equation), P_{ij} is the production of $\overline{u_i u_j}$ (which does not require modeling), K and ϵ are the turbulent kinetic energy and its rate of dissipation, and γ, C_1 and C_s are further empirical constants.

It is seen that :

- The closure assumptions are complicated, in particular the pressure-strain correlation closure.
- The K and ϵ equations are required to complete the closure schemes.
- Isotropic dissipation of $\overline{u_i u_j}$ was assumed.
- The task of finding the correct empirical constants is much more complex than for the K-ϵ model.

Reynolds stress models present the advantages of greater independence with respect to analogies between laminar and turbulent diffusion (ν_t and gradient-diffusion hypotheses), and a more complete account of turbulent transport processes. The main disadvantages are the large number of equations to be solved simultaneously, the dependence on simpler equations (in particular the ϵ equation), and the large number of empirical constants that must be determined.

2.9.6 The *K*-ϵ-*A* model.

Rodi (1980) suggested that in many turbulent flows, the transport of $\overline{u_i u_j}$ is small if compared with the change of $\overline{u_i u_j}$ itself. If this approximation is reasonable, the transport of $\overline{u_i u_j}$ is proportional to the transport of K, the proportionality factor being $\overline{u_i u_j}/K$. Thus, in this simplified Reynolds-stress model

$$\frac{D}{Dt}\overline{u_i u_j} - Diff(\overline{u_i u_j}) \rightarrow \frac{\overline{u_i u_j}}{K}\left(\frac{DK}{Dt} - Diff(K)\right) = \frac{\overline{u_i u_j}}{K}(\mathcal{P} - \epsilon)$$

using the K equation. Substituting this expression into the modeled $\overline{u_i u_j}$ transport equation, $\overline{u_i u_j}$ can be determined in terms of K and ϵ by means of the following equations (Rodi, 1980; Chen & Chang, 1987)

$$\frac{\overline{u_i u_j}}{K}(\mathcal{P} - \epsilon) = P_{ij} - C_1\frac{\epsilon}{K}(\overline{u_i u_j} - \frac{2}{3}\delta_{ij}K) - \gamma(P_{ij} - \frac{2}{3}\delta_{ij}\mathcal{P}) - \frac{2}{3}\epsilon\delta_{ij}$$

$$\overline{u_i u_j} = K\left\{\frac{\frac{2}{3}(C_1 - 1)\epsilon\delta_{ij} + (1-\gamma)P_{ij} + \frac{2}{3}\gamma\mathcal{P}\delta_{ij}}{\mathcal{P} + (C_1 - 1)\epsilon}\right\} \tag{2.32}$$

This (algebraic) equation, together with (2.31) and (2.30) constitutes the so-called *K*-ϵ-*A* model (*A* stands for "algebraic").

The *K*-ϵ-*A* model is simpler than the full Reynolds stress model, while retaining part of its qualities. At present, the eventual superiority of the *K*-ϵ-*A* with respect to the standard *K*-ϵ model is still an open question (*e.g.* Chen & Chang, 1987).

2.9.7 Practical considerations.

> *"Indeed, whether it be to predict the behaviour of a troublesome boundary-layer flow or to attack a recirculating flow that no one has ever considered calculating before, it is likely that a two-equation model of turbulence provides the best starting point; and perhaps the best finishing point as well."*
>
> Launder & Spalding, 1972

For the purposes of the present work, the *K*-ϵ model was chosen as the basis for the representation of turbulence because :

- It has been successfully employed in engineering problems involving complex turbulent flows (Rodi, 1980; Reynolds & Cebeci, 1978), such as:

open and closed channel flows, secondary flow in square ducts, flow in a square cavity, recirculating flows past obstacles, jet and wake dispersion, etc.

• It is neither too simple nor too complex to implement numerically (in terms of discretizations, boundary conditions, etc.) , and thus is a good approach for investigating new problems on which existing knowledge is scarce.

• It takes into account the following turbulent transport processes: time-variation, convection, production and dissipation (though not redistribution of Reynolds stresses).

2.10 Boundary conditions for K and ϵ.

2.10.1 Wall boundary conditions.

Boundary conditions for K and ϵ near a solid boundary are specified using universal laws which express the local state of turbulence in terms of the mean tangential velocity at a certain distance from the wall (Rodi, 1980).

Consider a region close to a flat wall, coincident with the x-axis, with the y axis pointing away from it. For small y, the wall may be considered infinite. If the time variation, convection and pressure gradient are small compared with the cross-stream shear stress (this is called the thin shear-layer approximation), the streamwise momentum equation is

$$-\frac{\partial \overline{uv}}{\partial y}+\nu\frac{\partial^2 U}{\partial y^2}=\frac{1}{\rho}\frac{\partial \tau}{\partial y} \tag{2.33}$$

and neglecting transport processes in the K equation

$$-\overline{uv}\frac{\partial U}{\partial y}-\epsilon=0 \tag{2.34}$$

In such conditions, $\tau=\tau_w$ and $\mathcal{P}-\epsilon=0$, *i.e.* close to the wall there is a region of constant stress, in which production of turbulence equals dissipation. An important parameter is the friction velocity, $u_*=(|\tau_w|/\rho)^{1/2}$; with this velocity it is possible to define a dimensionless velocity $U^+=U/u_*$, and a dimensionless distance $y^+=u_* y/\nu$.

Wall flows in steady, constant-stress turbulent boundary layers have been extensively studied (Hinze, 1975; Townsend, 1976); figure 2.4 (after Bradshaw, 1978) shows the typical profile of U^+ *vs* y^+ for those flows. The inner-layer is the region inside the boundary layer where the assumptions $\tau = \tau_w$ and $\mathcal{P} = \epsilon$ are valid. The properties of the flow in this region depend on the relative importance of the viscous and Reynolds shear stresses in (2.33), the roughness of the wall, and the pressure gradient (if it is not negligible).

Close to the wall, the Reynolds stress should be smaller than the viscous stress; hence, integrating (2.33) in y

$$\frac{U}{u_*} = \frac{u_* y}{\nu} \rightarrow U^+ = y^+$$

and thus the name "linear sublayer" . This equation is valid for $y^+ < 3$ approximately (figure 2.4).

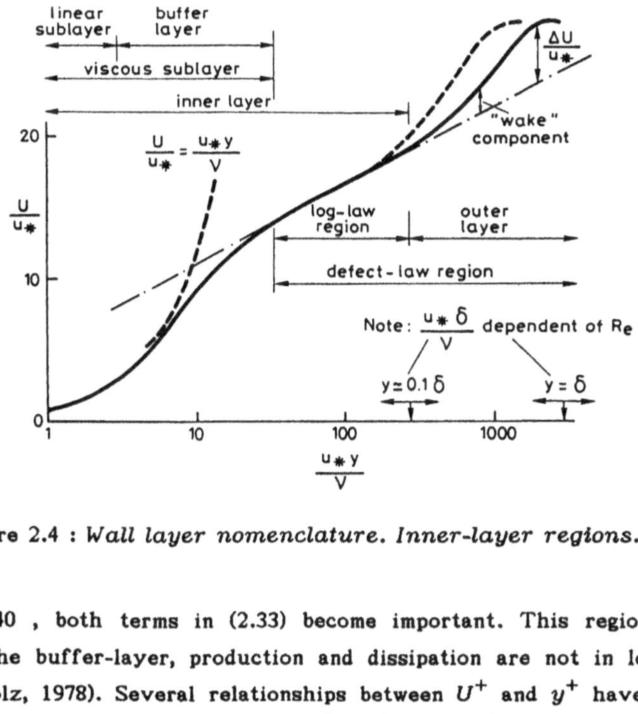

Figure 2.4 : *Wall layer nomenclature. Inner-layer regions.*

For $3 < y^+ < 40$, both terms in (2.33) become important. This region is called the "buffer-layer". In the buffer-layer, production and dissipation are not in local equilibrium (Hinze, 1975; Fernholz, 1978). Several relationships between U^+ and y^+ have been proposed for this layer. These relationships reduce to $U^+ = y^+$ when $y^+ \rightarrow 0$, and to the logarithmic law-of-the-wall for large y^+, but none can be considered entirely satisfactory (Hinze, 1975).

For $y^+ > 40$, the viscous contribution to τ is negligible, if compared with the Reynolds shear stress. In such circumstances, turbulence is in a state of local equilibrium. If the wall is smooth y is the only relevant length scale (Bradshaw, 1978; Townsend, 1976). The properties of the flow are determined by y, τ_w, ρ, and ν. Thus, using dimensional analysis

$$\frac{U}{u_*} = f_1\left(\frac{u_* y}{\nu}\right) = f_1(y^+) \tag{2.35}$$

$$\frac{\partial U}{\partial y} = \frac{u_*}{y} f_2\left(\frac{u_* y}{\nu}\right) = \frac{u_*}{y} f_2(y^+) \tag{2.36}$$

$$\frac{-\overline{uv}}{u_*^2} = f_3(\frac{u_* y}{\nu}) = f_3(y^+) \tag{2.37}$$

Considering the asymptotic behavior of (2.35)–(2.37) as $y^+ \to \infty$, it is immediate that $f_3(y^+) \to 1$; since $\partial U/\partial y$ should become independent of y^+, $f_2(y^+)$ tends to a constant, say $\frac{1}{\kappa}$. Thus

$$\frac{\partial U}{\partial y} = \frac{u_*}{\kappa y} \quad ; \quad \frac{U}{u_*} = \frac{1}{\kappa} ln(y) + A$$

Requiring compatibility with (2.35)

$$\frac{U}{u_*} = \frac{1}{\kappa} ln(\frac{u_* y}{\nu}) + C \quad ; \quad \frac{U}{u_*} = \frac{1}{\kappa} ln(E\frac{u_* y}{\nu})$$

This is the logarithmic law-of-the-wall, which is used to connect "outer velocities" (U) with wall conditions (u_*). For smooth walls, $\kappa \simeq 0.41$ and $E \simeq 9.0$ (Rodi, 1980).

Using the simplified K equation, a boundary condition for ϵ is found at once

$$\epsilon = -\overline{uv}\frac{\partial U}{\partial y} = \frac{u_*^3}{\kappa y} \tag{2.38}$$

Invoking the eddy-viscosity concept, the result of the mixing length theory is recovered

$$\nu_t = \frac{-\overline{uv}}{\frac{\partial U}{\partial y}} = \kappa u_* y$$

i.e. the eddy-viscosity is proportional to the distance from the wall. Expressing ν_t in terms of K and ϵ, a boundary condition for K is found

$$K = \frac{u_*^2}{(C_d)^{1/2}} \tag{2.39}$$

Thus, wall boundary conditions for K and ϵ may be derived by *i)* computing u_* as a function of U using the law-of-the-wall, and *ii)* computing K and ϵ using (2.39) and (2.38).

If the pressure gradient is not negligible, it is found by integration of the streamwise momentum equation that $\tau = \tau_w + \alpha y$, with $\alpha = dP/dx$. It can be shown that in such conditions, the law of the wall is changed to (Fernholz, 1978)

$$\frac{U}{u_*} = \frac{1}{\kappa} ln(\frac{u_* y}{\nu}E) + \frac{2}{\kappa}[ln(\frac{2}{(1+z)^{1/2}+1}) + (1+z)^{1/2} - 1]$$

with $z = \frac{\alpha y}{\tau_w}$.

If the wall is not impermeable, and fluid is injected into the domain with a velocity V_{wall} normal to the wall, the shear stress and the modified law-of-the-wall are (Fernholz, 1978)

$$\tau = \tau_w + \rho U V_{wall}$$

$$2\frac{u_*}{V_{wall}}[(1+\frac{UV_{wall}}{u_*^2})^{1/2}-1] = \frac{1}{\kappa}ln(\frac{u_* y}{\nu}E)$$

If the wall is not smooth, the influence of roughness is determined by the dimensionless roughness length (or roughness Reynolds number) $K^+ = u_* K_r/\nu$, with $K_r =$ *typical dimension of roughness elements.* Under the some assumptions as before (Hinze, 1975; Townsend, 1976)

$$\frac{U}{u_*} = f(\frac{u_* y}{\nu}, \frac{u_* K_r}{\nu})$$

$$U^+ = f(y^+, K^+)$$

The function f should become independent of K^+ as $K^+ \rightarrow 0$, and independent of viscosity as $y^+, K^+ \rightarrow \infty$. Geometrical shapes and distribution patterns of the roughness elements are extremely varied, and empirical correlations for the evaluation of f are not available. However, some investigation has been done on the properties of boundary layers over sand-grain roughness. Fernholz (1978) distinguishes three cases :

1) Hydraulically smooth roughness: $0 \leq K^+ \leq 5$
The roughness elements do not affect the logarithmic sublayer significantly. This corresponds to the case $U^+ = f(y^+)$, *i.e.* to the usual logarithmic law-of-the-wall.

2) Transient roughness: $5 \leq K^+ \leq 70$
The velocity profile is affected by the roughness of the wall. This effect is taken into account by substituting the constant C in $U^+ = \kappa^{-1} ln(y^+) + C$ by a function of K^+ . The appropriate function is (Fernholz, 1978)

$$U^+ = \frac{1}{\kappa}ln(y^+) + C - \frac{1}{\kappa}ln(K^+ + 3.30) + 2.92$$

with $C = 5.0$ to 5.2 .

3) Fully developed roughness: $K^+ \geq 70$
In this case, the velocity distribution should be independent of viscosity. The law-of-the-wall takes the following form

$$U^{+} = \frac{1}{K} ln(\frac{y}{z_0})$$

Here, z_0 is chosen in such a way that it absorbs any dimensionless constants.

The law-of-the-wall is also applicable in unsteady turbulent boundary layers, if local acceleration is moderate, and the free-stream velocity sufficiently high. This conclusion, which is discussed in Sumer et al. (1986) and Cousteix et al. (1977), is of great practical importance in modeling unsteady turbulent flows with the K-ϵ model.

2.10.2 Planes and axes of symmetry.

At symmetry planes or lines, the normal gradients of K and ϵ should vanish, as well as the Reynolds shear stresses (Launder, 1989; Rodi, 1980).

2.10.3 Turbulent non-turbulent interfaces.

At a turbulent-nonturbulent flow transition boundary, velocities and scalar quantities are equal to their free-stream values. Two cases can be distinguished: either the free-stream is turbulence-free, or it is a homogeneous shear-free turbulent flow; the turbulent-nonturbulent boundary is assumed to be fixed in space (this is only a crude approximation; such boundaries are known to be unsteady and highly complicated (intermittent)—Tennekes & Lumley, 1972).

In the first case, all turbulence stresses and fluxes and the rate of dissipation ϵ are zero at the transition boundary. This poses problems to the standard K-ϵ model, which must be modified by *ad-hoc* assumptions (Jones & Launder, 1972) in order to remove the singularities in the ϵ-equation.

In the second case, Rodi (1980) suggests that the boundary conditions should be determined using the K-ϵ equations with lateral convection, diffusion and production neglected.

2.10.4 Free-surface boundary conditions.

Free-surface boundary conditions for turbulence-related variables are extremely difficult to prescribe. Due to air entrainment, it is even difficult to define the free-surface precisely.

There are two main processes which produce turbulence near to a free-surface: air friction, and internal recirculation. The former is important in nonbreaking flows, such as channel and tidal flows. In a breaking free-surface flow (which is the case of interest), both processes are important, but internal recirculation may be considered dominant (Mizuguchi, 1986; Svendsen, 1987).

If the free-surface has small curvature, it can be (conceptually) substituted by a rigid-lid (Celik et al. , 1987); then, boundary conditions appropriate for a solid wall are employed. Clearly, this solution is too simple to be adopted in complex free-surface turbulent flows. Rodi (1980) presented a solution for the case of channel flow. If wind-induced shear stress with the surface is small, he proposes to treat the free-surface as a symmetry plane; when a shear-layer is created by forces near the surface, he proposes treating the free-surface as a moving wall. The mentioned author proposes another idea to complement his formulation, recognizing that the presence of the free-surface should reduce the length scale of turbulence and the eddy-viscosity coefficient at the boundary. Thus, the final proposal is (Rodi, 1980)

$$K_{surf}(C_d)^{1/2} > u_{*surf}^2 \quad \rightarrow \quad \textit{Symmetry condition for } K$$

$$K = \frac{u_{*surf}^2}{(C_d)^{1/2}} \quad \rightarrow \quad \textit{Otherwise}$$

$$\epsilon = \frac{K^{3/2} C_d^{3/4}}{\kappa[y + a_* h(1 - \frac{u_{*surf}^2}{K_{surf}(C_d)^{1/2}})]}$$

In these equations, the subscript $(\;)_{surf}$ refers to conditions at the free-surface, a_* is a further empirical constant and h is the total thickness of the shear layer.

This solution is not valid in a turbulent recirculating (breaking) flow, where turbulence is mainly produced by internal recirculation, and the free-surface may be highly curved; a simple prescription for this case is proposed in § 5.7 .

Chapter 3

MATHEMATICAL MODELING OF BREAKING SHALLOW WATER WAVES PROPOSED METHODOLOGY

3.1 Introduction.

In this chapter, the essential aspects of wave transformation in shallow water are reviewed, with special emphasis on the breaking phenomenon in surf-zones.

From a physical viewpoint, surf-zones are characterized by the irreversible transformation of organized wave motion of the incident short (wind-generated) waves, into motions of different types and scales. The initial wave motion is almost irrotational, and while the free-surface remains simply-connected, vorticity can only spread into the body of the fluid from the boundary layers by the action of viscosity. After the waves break, intense vorticity is generated by the folding of the free-surface onto itself. In consequence, strong levels of turbulence and energy dissipation arise. These are accompanied by the appearance of other types of flows. The main types of flow found in the surf-zone are summarized in table 3.1 , by showing their typical space and time scales, as well as their role in surf-zone hydrodynamics.

Table 3.1 : Typical space and time scales for the three main flow modes found in surf-zones.

Flow mode	Time scale	Space scale	Physical role
Turbulence	$10^{-3}-1\ s$	$10^{-4}-1\ m$	Dissipates and redistributes energy
Wind waves	$1-20\ s$	$1-100\ m$	Driving agent (energy income)
Low-frequency motion	$\geq 20\ s$	$100-1000\ m$	Changes bottom topography (plus other effects)

The main features associated with the transformation of an incident wave train across a surf-zone are illustrated in figure 3.1 . In this figure, SLW denotes the still water level, MLW denotes the mean water level (in the presence of waves), Br denotes the breaking point (which is not unique, because of wave irregularity), and R is the aerated, strongly turbulent region in front of broken waves, usually called "wave-roller". Also, the surf-zone is divided into three regions, which have been called "outer region", "inner-region", and "run-up region" (Peregrine & Svendsen, 1978).

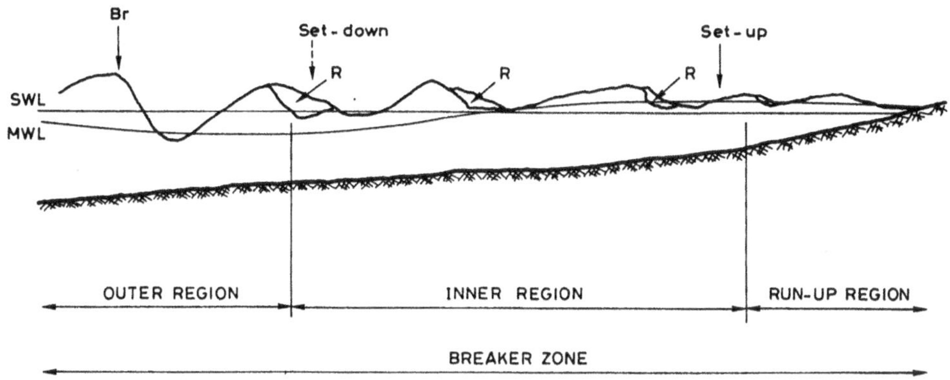

Figure 3.1 : *Wave transformation across a typical surf-zone.*

In the outer region, the incident waves become unstable, quickly change their form, and break. In the inner region, the broken waves propagate with a very gradual change of form, and resemble moving hydraulic jumps (bores). The ordered oscillatory wave motion is transformed into well developed turbulent motion and low-frequency motion. The essential features of turbulence dynamics in inner region waves are illustrated in figure 3.2 (adapted from Svendsen, 1987). The prominent characteristic of inner-region waves is a highly turbulent, recirculating mass of air and water in their fronts, called "wave roller". In the run-up region, there is no wave roller and only a thin turbulent lamina of fluid persists.

The study of breaking wave propagation and decay can in principle be carried on by solving the instantaneous wave motion, or by giving a global description of waves in terms of spatial variation of their integral properties, treating the waves locally as a uniform wave train (Battjes, 1988).

In principle the two methodologies are equivalent. Using the instantaneous description, the global properties and the low-frequency motion can be determined by appropriate averages of the solution. Using the global description, it is assumed that the mechanisms of the flow in breaking waves are too complex to be described in detail. Then, these mechanisms are expressed in terms of the mass, momentum and energy conservation principles averaged over a typical wave period (Mei, 1983; Phillips, 1977). The actual wave motion is not represented with this methodology, and the driving influence of the incident wave field must be specified as an external input of semi-empirical nature. This approach is the basis of most nearshore circulation models used in practical applications.

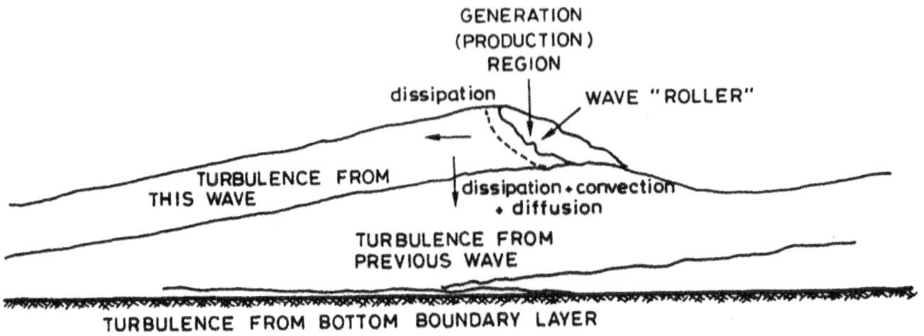

Figure 3.2 : *Turbulence dynamics in inner region breaking waves (adapted from Svendsen,* 1987).

The global description has the advantages of economy and simplicity, but is based on some assumptions that are not valid inside a surf-zone, such as considering the waves to be irrotational, purely progressive, nearly plane and of permanent form. These assumptions are combined with empirical models representing turbulent mixing, with which they are not compatible. It is apparent that the quality of the global wave approach could be greatly improved through a better understanding of the details of the motion in breaking waves.

The present work is concerned with the latter type of methodology, dealing with the phenomena that take place in time scales comparable to a wave period. These phenomena are: the wave deformation at the onset of breaking; the production, dissipation and diffusion of vorticity and turbulence; wave energy dissipation and associated height decay; and wave-wave interactions.

The physical processes involved in surf-zone wave transformation are analyzed in § 3.2. Then, the mathematical descriptions that have been proposed are considered (§ 3.3). All these mathematical descriptions have important limitations. Consequently, it is concluded that a more general formulation, compatible with the physical processes, is needed. A new methodology is proposed for the numerical simulation of two-dimensional wave motion in the vertical plane, whose basic elements are introduced in § 3.6. Experimental investigations with special relevance for the numerical investigations performed in the present work are reviewed in §§ 3.4 and 3.5 .

3.2 Physical processes.

3.2.1 Wave breaking criteria.

A great deal of the research on wave breaking phenomena has consisted of experiments to determine when, where and what type of breaker results for given offshore characteristics and beach slope (Iribarren & Nogales, 1949; Galvin, 1968, 1972; Battjes, 1974; Battjes, 1988). From these investigations it is concluded that the qualitative aspects of wave breaking depend on two parameters: the beach slope m, and the initial wave steepness H_0/λ_0 (H_0=deep water wave height; λ_0=deep water wavelength). The type of breaker, degree of reflection and run-up are functions of the single dimensionless parameter I_r, known as the Iribarren parameter (Iribarren & Nogales, 1949; Battjes, 1974)

$$I_r = \frac{m}{(H_0/\lambda_0)^{1/2}}$$

Another equivalent parameter is $\chi_0 = a_0\omega^2/gm^2$ (a_0 = deep water wave amplitude; ω = wave angular frequency; g = acceleration due to gravity) (Galvin, 1968). The local parameter $\chi = a\omega^2/gm^2$ can be used as a theoretical criterion for the onset of breaking of periodic waves on a slope. Breaking occurs when $\chi_{br} \simeq 1$ (Carrier & Greenspan, 1958; Munk & Wimbush, 1969). Many other breaking criteria have been proposed. These criteria specify either the limiting wave steepness $(H/\lambda)_{br}$, or the breaker index $(H/D)_{br}$ (D = mean water depth, in the presence of waves). The breaking criterion obtained by McCowan (1894) for the solitary wave

$$(\frac{H}{D})_{br} = 0.78$$

is found to be appropriate for estimating the breaker height in mild-sloped surf-zones.

The main causes of breaking are disturbances, instabilities and wave evolution (Peregrine, 1988). Disturbances that can induce breaking are wind, a change of water depth, and fixed or moving obstacles. For this work's purposes, the two relevant types of disturbances are changes of water depth, and other waves (which somewhat behave as moving obstacles in wave-wave interactions). Instabilities may be an important cause of deep-water breaking (Benjamin & Feir, 1967; Longuet-Higgins, 1978a, 1978b), and certain types of instabilities also occur in shallow water waves (Tanaka, 1986). Wave evolution can also be an important cause of breaking, especially when the waves are close to the limiting steepness and/or do not have a steady form initially.

The sequence *disturbances* $\Rightarrow$ *wave evolution* $\Rightarrow$ *breaking* is fundamental for understanding the initial stages of the breaking process. This sequence is inhibited when the shallow water steepening and dispersive effects are in equilibrium (§ 3.2.2). When the effect of disturbances becomes prominent, this equilibrium is broken, the sequence is activated, and wave overturning is the result (§ 3.2.3).

3.2.2 Shallow water steepening *vs* dispersion.

In shallow water, the front face of any wave of elevation propagating on water of uniform depth will steepen and break, because the higher parts of the wave travel faster (Lamb, 1932; Whitham, 1974). The importance of this process is measured by the dimensionless parameter $\epsilon_w = H/D$ ($H=$ wave height, $D=$ water depth). As a result of shallow-water steepening, the vertical acceleration of the water particles becomes important in the steep front of the waves, and its effect on the pressure can no longer be neglected. The influence of vertical acceleration tends to counteract shallow-water steepening. This is called the dispersive effect, which is measured by the dimensionless parameter μ_w^2, where $\mu_w =$ measure of maximum gradients (*e.g.* $D\partial/\partial x$).

If the nonlinear steepening and the dispersive processes are in balance, then waves of finite amplitude and permanent form are possible. This balance is expressed in terms of the Ursell number (Ursell, 1953)

$$U_r = \frac{H}{D\mu_w^2}$$

Often μ_w is replaced by D/λ ($\lambda =$wavelength). However, this is not strictly correct for shallow water waves, since long-wavelength waves are similar to a train of solitary waves with long flat troughs between crests (hence their "effective" length is smaller than λ).

As the incident waves travel into shallower water, shallow-water steepening becomes predominant over dispersion. The wave fronts become steeper and ultimately break. After the breaking has started, the wave front is stabilized by turbulence. If the water depth increases, such as occurs when waves pass over a submerged bar, the breaking may stop if for the new depth $U_r \sim \mathcal{O}(1)$. An undular bore results (Peregrine, 1983; Battjes, 1988).

3.2.3 Wave overturning.

The first relevant details of wave overturning were obtained by the numerical approach introduced by Longuet-Higgins & Cokelet (1976), which has followed and refined by others. An example of successive profiles for an overturning wave are represented in figure 3.3 (after Peregrine, 1983).

A detailed study of overturning solutions (Peregrine et al. , 1980; Peregrine, 1983) reveals that the main features of the overturning motion are: water-particle velocities exceed the wave phase celerity; water-particle accelerations exceed g in a thin region on the front of the wave; there is an extensive region on and beneath the back slope of the wave where the water-particle accelerations are small (see figure 3.4). The significant feature of wave overturning is a strong convergence of water towards the (spilling or plunging) jet.

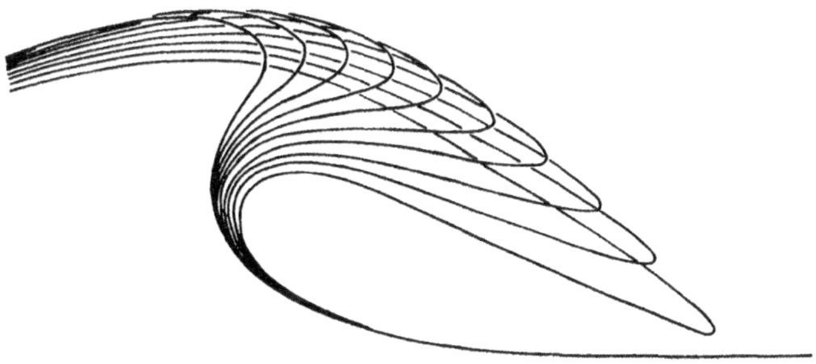

Figure 3.3 : *Sequence of computed wave profiles for a periodic wave after an "impulsive" depth perturbation.*

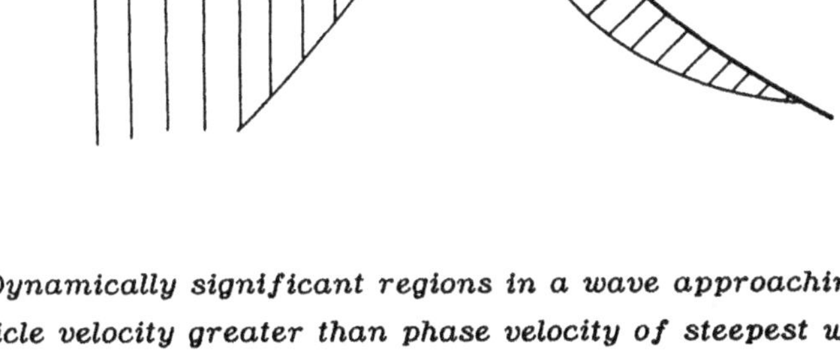

Figure 3.4 : *Dynamically significant regions in a wave approaching breaking:*

i) particle velocity greater than phase velocity of steepest wave;

ii) water accelerations greater than gravity;

iii) water accelerations less than 1/3 gravity.

(*After Peregrine et al.* , 1980)

3.2.4 Breaking wave propagation and decay.

When the plunging jet of a breaking wave touches the water in front of it, the wave dynamics change abruptly. The flow around the trapped air forms a vortex around which there is considerable circulation. Vorticity is impulsively generated, and so is turbulence. The sheet of water pushed by the plunging jet continues to fall down, repeating a cycle of splashing and vortex formation (Jansen, 1986; Peregrine, 1988). This cycle appears to be of a deterministic nature, even though it is obscured by drops, bubbles, and the growing level of turbulence (Peregrine, 1983, 1988). After this initial cycle of splashing and vortex formation, the flow becomes fully turbulent. The broken wave is gradually converted in a propagating spilling breaker or a turbulent bore depending on the initial type of breaker and the local water depth.

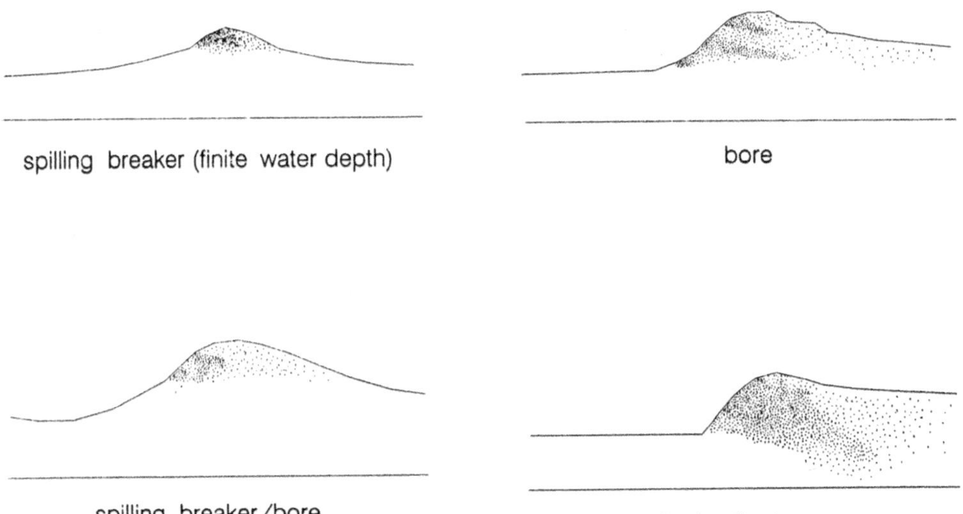

Figure 3.5 : *Categories ("spectrum") for inner-zone breaking waves (Adapted from Peregrine & Svendsen,* 1978).

Peregrine & Svendsen (1978) proposed a number of qualitative categories (which they call "spectrum") for classifying well-developed breaking waves. For the purposes of the present work, the categories of interest are "spilling breaker (finite water depth)", "spilling breaker/bore" , "bore" and "hydraulic jump". These categories are sketched in figure 3.5 .

3.2.5 Other physical effects.

Air

Air entrainment during the breaking process results in a variation of the density of the air-water mixture, non-negligible compressibility in the strongly aerated regions and complex turbulence-generating processes. It is generally accepted that the variation of density has a small direct influence on the dynamics (Peregrine, 1983; Battjes, 1988). The equation for the density of an air-water mixture is

$$\rho = \alpha_m \rho_{air} + (1 - \alpha_m)\rho_{water}$$

where α_m is the concentration of air. An estimate for α_m is provided by experimental studies of the roller in steady hydraulic jumps. Rajaratnam (1967) and Resh et al. (1976) give $0.1 \leq \alpha_m \leq 0.2$. Since $\rho_{air} \simeq 10^{-3}\rho_{water}$

$$\rho \simeq (1 - \alpha_m)\rho_{water}$$

Hence, the direct dynamic influence of air entrainment is small, but not entirely negligible in strongly aerated regions (*i.e.* the "wave roller").

The air-water mixture is not incompressible. Limited compressibility can be introduced by replacing the continuity equation $\nabla \cdot \vec{U} = 0$ by (Nichols et al. , 1980)

$$\frac{1}{\rho c^2}\frac{\partial P}{\partial t} + \nabla \cdot \vec{U} = 0$$

where c is the speed of sound in the fluid (mixture), ρ is the water density, $\vec{U}$ is the mean velocity and P is the mean pressure. It is known that c can be lowered by an order of magnitude due to air entrainment (Peregrine, 1983).

Nadaoka (1986) found experimentally that the spatial distribution of the entrained air bubbles under periodic breaking waves (measured by the drop-out index of the signal from the velocity measurement apparatus) is strongly correlated with the turbulent kinetic energy. However, typical values for the air concentration α_m and sound speed c are not available from his results. The determination of α_m and c in an unsteady breaking flow would require additional equations describing creation, destruction and transport of the air concentration, plus a constitutive relationship of the type $c = c(\alpha_m)$, with turbulence effects. Such equations are not established, and very little is known about the physical properties of the air/water mixture. In the present work no account is taken of these complexities.

Surface tension

Surface tension (capillary waves) plays an important role in sustaining wind stress on the sea. It is known that surface tension inhibits the formation of a plunging jet and air entrainment on steep waves less than 10 cm high. In the present work surface tension effects are neglected.

Drops and bubbles

The visual appearance of breaking waves is a result of the combination air/surface tension/drops and bubbles/turbulence. The importance of drops and bubbles in generating turbulence is not known (Peregrine, 1988).

From the above discussion, it is concluded that the effects of air in the breaking process are more pronounced if the relative velocity is high and the plunging jet small. Otherwise there can be considerable spray formation and air entrainment. The incompressibility assumption is a reasonable starting point, but has real limitations. Consequently, the model herein developed is expected to give more realistic results for a plunging breaker evolving to a moving hydraulic jump (where turbulence production due to internal recirculation outweighs turbulence production by air friction) than for a very gradual spilling breaker (where the region with strong turbulence remains close to the wave crest).

3.3 Mathematical descriptions.

The process of wave transformation across a surf-zone is governed by the NS equations. A direct solution of all details of the flow using these equations is at present unfeasible because the problem is highly nonlinear, the free-surface can assume very complicated configurations, and the flow is turbulent. To render the problem tractable, simplifying assumptions have to be introduced. On basis of such assumptions, either analytical or numerical descriptions of the wave motion can be obtained. Wave theories are examples of analytical descriptions. The shallow-water equations (*abbr.* SWE´s), Boussinesq SWE´s and numerical models based on the primitive equations are examples of numerical descriptions. In this section, these mathematical methods are reviewed, with emphasis on the nature of their basic assumptions, their range of validity and limitations, and the role they play in the present work.

3.3.1 Wave theories. Range of validity.

Most wave theories are based on the following assumptions: the motion is two-dimensional; the fluid is ideal and the flow irrotational; the waves propagate without change of form, and the free-surface is a single-valued function of x and t; and the bottom is horizontal. These theories provide analytical expressions for the instantaneous and integral properties. The wave problem is greatly simplified if the solution is obtained in a reference frame moving with the wave celerity c. In this reference frame the motion is steady, and the solution unique. Stokes (1847) demonstrated that the wave celerity must be specified by some physical consideration. The steady wave problem is schematized in figure 3.6, where x and y are the horizontal and vertical coordinates; η is the free-surface elevation; λ is the wavelength; D is the mean water depth; d is the undisturbed water depth; h is the water depth under the troughs; u_w and v_w are the velocities in the moving reference frame; and $U_w = u_w + c$, $V_w = v_w$ are the velocities in the fixed reference frame.

If the is fluid incompressible and the motion irrotational, there exists a stream function $\psi(x,y)$ such that $u_w = \partial\psi/\partial y$, $v_w = -\partial\psi/\partial x$ and $\nabla^2\psi = 0$ throughout the fluid. The BC's for the governing (Laplace) equation are: on the bottom $\psi(x,0) = 0$; on the free-surface $\psi(x,\eta(x)) = Q$, and in addition the pressure is a constant, $= 0$ say, from which it follows that $1/2(\nabla\psi)^2 + g\eta = R$. The assumed periodicity in x (or conditions for $|x| \rightarrow \infty$) provides the lateral boundary conditions. The properties of the wave system are completely defined by the three quantities

$$Q = \int_0^\eta u_w dy \tag{3.1}$$

$$R = \varphi_w + gy + \frac{1}{2}(u_w^2 + v_w^2) \tag{3.2}$$

$$S = \int_0^\eta (\varphi_w + u_w^2) dy \tag{3.3}$$

which are the volume flow rate per unit span, the total energy per unit mass, and the total momentum flux per unit mass per unit span, respectively (Benjamin & Lighthill, 1954). These quantities are invariant for a wave system without friction or other losses. Solutions to the wave problem have been obtained by either perturbation or numerical methods (Schwartz & Fenton, 1982). Perturbation methods (*e.g.* Stokes, 1847; Laitone, 1960; Fenton, 1972, 1979, 1985; Schwartz, 1974; Cokelet, 1977) have the advantage of providing analytical expressions for the wave quantities. However, they usually have a limited range of validity and may require convergence improvement techniques. Numerical methods (*e.g.* Dean, 1965; Rienecker & Fenton, 1981) present the advantage of high accuracy, but are less manageable and informative.

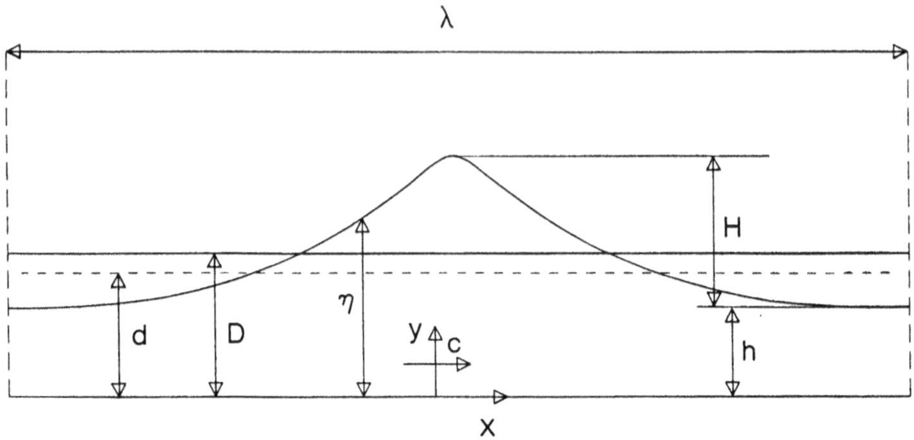

Figure 3.6 : *Definition sketch for the water wave problem.*

Perturbation methods consist of expansions in terms of some small parameter. The type of expansion and consequent range of validity depends on which parameter is chosen as expansion parameter. In engineering applications, two types of expansions have been employed: Stokes ' expansions and shallow-water expansions. In Stokes ' expansions the perturbation parameter is $\epsilon_w = Hk/2$, in which $k = 2\pi/\lambda$ is the wave number. The following expansion is assumed for ψ (Fenton, 1985)

$$\frac{k\psi}{\overline{u}} = -ky + \sum_{i=1}^{\infty}\sum_{j=1}^{i} f_{ij}(\epsilon_w)^i sinh(jky)cos(jkx) \tag{3.4}$$

where $\overline{u}$ is the mean velocity for a point always in the fluid. Perturbation series in ϵ_w are also assumed for the quantities in (3.4) and in the boundary conditions ($\overline{u}$, Q, k, η). After expanding and performing the necessary series manipulations, like- powers of ϵ_w are equated and the unknown coefficients are found. For shallow water waves, a different form for the perturbation series and a different expansion parameter are chosen. The appropriate form for ψ is the Rayleigh-Boussinesq series

$$\psi = -sin(y\mathfrak{D})f(x,0) \tag{3.5}$$

$$sin(y\mathfrak{D}) = y\frac{\partial}{\partial x} - \frac{y^3}{3!}\frac{\partial^3}{\partial x^3} + \cdots$$

and the solutions are the cnoidal and solitary waves. Expansions of up to the fifth order were obtained by several authors (see Fenton, 1985) for Stokes expansions, and by Fenton (1979) for shallow water expansions (cnoidal and solitary waves).

Both (3.4) and (3.5) satisfy the governing equation and the bottom BC automatically. Stokes-type expansions are not well suited for shallow water because the successive terms actually behave like $\epsilon_w / sinh^3(kD)$ (Schwartz & Fenton, 1982; Dean & Dalrymple, 1984). On the other hand, shallow water expansions do not work for deep water waves, because the nominal expansion parameter is $H/(mh)$, where m is the parameter of elliptic functions. For the long-wavelength limit of the solitary wave $m = 1$, but for shorter waves m becomes smaller and the expansion parameter large, deteriorating the convergence of the series. Thus, Stokes and shallow water expansions are complementary, and both have a limited range of validity. Other perturbation methods have been proposed, in particular the very high order solutions of Schwartz (1974) and Cokelet (1977), but these are of an inverse nature, and thus not convenient for practical applications.

Wave theories are incompatible with the physical processes of wave transformation in the surf-zone, for their underlying assumptions are too restrictive. They are useful outside the breaker area for explaining some aspects of the sequence *disturbances ⇒ wave evolution* (*e.g.* Longuet-Higgins & Fenton, 1974; Longuet-Higgins & Cokelet, 1976). In the present work, the main interest is in solitary wave theories, which are the limiting solutions of shallow water expansions for $D/\lambda \rightarrow 0$. These theories provide a good description of irrotational waves near the breaking point, and explain certain aspects of the difference between plunging and spilling breakers (Longuet-Higgins & Fenton, 1974). They find a direct application in the generation of initial and boundary conditions for the present numerical model, and are discussed in more detail in § 3.4.1 .

3.3.2 Shallow-water equations. Characteristics and discontinuities.

The nonlinear shallow-water equations (*abbr.* SWE's) provide a simple model for explaining wave steepening and breaking. For our purposes, it will be assumed that the fluid is ideal and the bottom horizontal, and that the waves are very long, *i.e.* $kD \ll 1$. The horizontal velocity u_w is assumed to be uniform in y. The vertical velocity and acceleration are assumed to be negligible; this implies a hydrostatic pressure distribution, *i.e.* any dispersive effects are neglected. The 1-D SWE's are (Abbott, 1979)

$$\frac{\partial \eta}{\partial t} + \frac{\partial}{\partial x}(u_w \eta) = 0$$

$$\frac{\partial u_w}{\partial t} + u_w \frac{\partial u_w}{\partial x} + g \frac{\partial \eta}{\partial x} = 0$$

Introducing $c = (g\eta)^{1/2}$, these equations can be put into the characteristic form (Abbott, 1979)

$$\frac{\partial}{\partial t}(u_w \pm 2c) + (u_w \pm c)\frac{\partial}{\partial x}(u_w \pm 2c) = 0$$

The quantities $J_{\pm} = u_w \pm 2c$ (Riemann invariants) are constant along the lines $\frac{dx}{dt} = u_w \pm c$ (characteristic lines) . Any excess in elevation upstream induces a pile-up of characteristics, steepening the wave front until it becomes vertical. All waves carrying an excess in elevation will break, unless there is bottom friction (real-fluid effects) or dispersion. Hence, this theory cannot explain the existence of waves of permanent form. When a discontinuity is formed, the governing equations are replaced by the following transition relations (shock, or Rankine-Hugoniot conditions—see Abbott, 1979 , or Whitham, 1974)

$$-C[\eta]_1^2 + [u_w \eta]_1^2 = 0$$
$$-C[u_w \eta]_1^2 + [u_w^2 \eta + \frac{1}{2} g \eta^2]_1^2 = 0$$

where $[\ ]_1^2$ denotes the difference of the quantities within brackets on either side of the discontinuity, and C is the celerity at which the discontinuity propagates (§ 3.4.2). These transition relations are derived from the assumption that mass and momentum are conserved across the discontinuity. However, energy must be dissipated.

Shallow-water theory is useful in explaining the tendency of long waves to break, ceases to be valid when the wave front becomes steep and the waves break, but is again useful in the inner region, where the broken waves are reasonably well described as turbulent bores (*i.e.* weak solutions of the SWE's). Turbulent bores and hydraulic jumps play an important role in the present work, and are discussed in more detail in § 3.4.2 .

3.3.3 Boussinesq-type shallow water equations.

The so-called Boussinesq-type SWE's describe finite-amplitude dispersive waves in shallow water. The effects of nonlinear steepening and dispersion are introduced by means of a formal expansion in terms of the two parameters, ϵ_w and μ_w. The following assumptions are introduced : the fluid is ideal, and the flow irrotational; the bottom is horizontal; $\epsilon_w \ll 1$, $\mu_w \ll 1$ and $\frac{\epsilon_w}{\mu_w} \sim \mathcal{O}(1)$, *i.e.* nonlinear steepening and dispersion effects are small and of the same order of magnitude. The velocity potential $\phi(x,y)$ satisfies $\nabla^2\phi = 0$ and $\partial\phi/\partial y = 0$ on $y = -D$ (in this section, η is measured from the mean water level for convenience). The small μ_w suggests an expansion of the Rayleigh-Boussinesq type for ϕ

$$\phi = cos(y\mathfrak{D}) f(x,t)$$

$$cos(y\mathfrak{D}) = 1 - \frac{y^2}{2!}\frac{\partial^2}{\partial x^2} + \frac{y^4}{4!}\frac{\partial^4}{\partial x^4} - \cdots$$

The two dimensionless parameters ϵ_w and μ_w are introduced by scaling the horizontal and vertical dimensions differently (Friedrichs, 1948; Whitham, 1974)

$$x \rightarrow x/\lambda \qquad y \rightarrow y/D \qquad \eta \rightarrow \eta/D$$

$$t \rightarrow \frac{(gD)^{1/2}}{\lambda} t \qquad \phi \rightarrow [\frac{H\lambda}{D}(gD)^{1/2}]^{-1}\phi$$

Inserting the expression for ϕ into the free-surface boundary conditions, the following equations are obtained, using the same symbols for the dimensionless variables (Whitham, 1974; Mei, 1983)

$$\frac{\partial \eta}{\partial t} + \frac{\partial}{\partial x}[(1+\epsilon_w \eta)u_w] + \mathcal{O}(\epsilon_w \mu_w^2, \mu_w^4) = 0 \tag{3.6}$$

$$\frac{\partial u_w}{\partial t} + \epsilon_w u_w \frac{\partial u_w}{\partial x} + \frac{\partial \eta}{\partial x} - \frac{1}{3}\mu_w^2 \frac{\partial^2 u_w}{\partial t \partial x^2} + \mathcal{O}(\epsilon_w \mu_w^2, \mu_w^4) = 0 \tag{3.7}$$

where u_w is the depth-integrated horizontal velocity

$$u_w = u_0 - \frac{\mu_w^2}{6}(1+\epsilon_w \eta)^2 \frac{\partial^2 u_0}{\partial x^2} + \mathcal{O}(\mu_w^4)$$

and u_0 is the velocity at the bottom. The dispersive effect is expressed by the term containing third order derivatives in (3.7), which is proportional to μ_w^2. To $\mathcal{O}(\mu_w^2)$ the horizontal velocity is uniform in y. From the expansion for ϕ it is inferred that the vertical velocity varies linearly from the bottom to the free-surface. When the dispersive term is neglected in (3.7), *i.e.* $\mu_w^2 \rightarrow 0$, the SWE's are obtained. When the nonlinear term is neglected in (3.6)–(3.7), *i.e.* $\epsilon_w \rightarrow 0$, linear waves with a dispersion relation $\omega^2 = gDk(1-\frac{1}{6}D^2k^2)$ are obtained (Miles, 1981; Whitham, 1974). This is the first approximation to the full dispersion relation for linear waves. The Boussinesq equations admit both left- and right-running waves. Assuming unidirectional propagation, the Korteweg-de Vries (*abbr.* KdV) equation (Korteweg & de Vries, 1895) is obtained (see *e.g.* Whitham, 1974)

$$\frac{\partial \eta}{\partial t} + (1+\frac{3}{2}\epsilon_w \eta)\frac{\partial \eta}{\partial x} + \frac{1}{6}\mu_w^2 \frac{\partial^3 \eta}{\partial x^3} + \mathcal{O}(\epsilon_w^2 + \mu_w^4) = 0$$

The Boussinesq and KdV equations are valid for small ϵ_w and μ_w^2 , and in particular for $\epsilon_w/\mu_w^2 \sim \mathcal{O}(1)$. Their essential feature is the ability for describing waves in which nonlinearity and dispersion are in perfect balance. The KdV equation admits periodic solutions (cnoidal waves) and the (first order) solitary wave. It is well known that unidirectional solitary waves of different amplitude pass through one another without any permanent change of shape and suffer only phase shifts. Solutions of nonlinear equations with these properties are called solitons (see *e.g.* Argyris & Haase (1987) for a survey of soliton theory, with numerical applications).

The Boussinesq equations are a better description to shallow water waves approaching the breaking point that the SWE's. However, due to the nature of the approximations, they can neither describe in detail the onset of breaking, nor be used inside the surf-zone.

3.3.4 Overturning wave models.

The main difficulties for a mathematical description of the overturning process are the complex structure of the flow before overturning, the quick deformation of the overturning wave, the complexity of the free-surface evolution, and the resulting impulsive generation of vorticity. Numerical simulations of wave overturning have been performed using either time-dependent irrotational flow models, or MAC-type models.

Time-dependent irrotational flow models are based on the solution of a Laplace equation for the velocity potential ϕ in a time-varying domain, whose configuration is updated numerically (*e.g.* Baker et al. , 1982). These models have been successfully used in studying the overturning process (Longuet-Higgins & Cokelet, 1976; Vinje & Brevig, 1981; Dold & Peregrine, 1984, 1986), the effect of disturbances (Cooker et al. , 1989), and instabilities (Tanaka et al. , 1987). The advantages of time-dependent potential flow models with respect to MAC-type models are greater economy and higher precision (for the same computational effort) when the flow is effectively irrotational. However they are not sufficiently general for modeling surf-zone waves, because they cannot represent vorticity. The computed solutions are only valid up to the moment when the plunging jet touches the water beneath it. It is impossible to describe the subsequent evolution of the broken wave. This severe limitation can only be overcome by considering more general alternatives, such as MAC-type models.

The MAC (Marker-and-cell) method was introduced (Welch et al. , 1966) for solving time-dependent incompressible and viscous flows involving free-surfaces. It consisted of a numerical algorithm for solving the (full) momentum and pressure Poisson equations for incompressible flows, together with a set of marker particles (massless tracers) for defining the location of the free-surfaces within the computational domain. The MAC method proved to have a considerable potential for fluid flow problems in general, and for water wave problems in particular, because

- The full incompressible flow equations are used. All restrictions arise only from the discrete approximations.
- Arbitrary free-surface configurations can be represented.
- Turbulence models can be incorporated, in order to represent turbulence effects.

These features of MAC made it the basis for many subsequent models for free-surface flow problems, which have been applied in dam-break problems (Welch et al. , 1966; Amsden & Harlow, 1970; Nichols et al. , 1980), solitary wave propagation and reflection (Chan & Street, 1970), ship waves and waves breaking over obstacles (Miyata, 1986). Nevertheless, MAC-type models also have limitations for application to surf-zone waves: they are computationally expensive, and the free-surface boundary conditions are difficult to apply without simplifications (§ 4.7.3). Another limitation is the incompressibility assumption, as was discussed in § 3.2.5 .

Turbulence transport models have been coupled with the basic MAC solution algorithm (*e.g.* Harlow & Nakayama, 1967, 1968; Amsden & Harlow, 1968; Daly & Harlow, 1970), but these models where different from those nowadays established, and no attempt was made to apply these extensions to complex turbulent free-surface flows.

3.4 Wave theories for very shallow water.

3.4.1 Solitary waves.

The solitary wave is a good approximation for irrotational waves in very shallow water, *i.e.* close to the breaking point (see *e.g.* Munk, 1949; Stiassnie & Peregrine, 1980). Solitary waves are nonlinear solutions which allow a precise check of the accuracy of numerical free-surface flow models. Also, they possess important interaction properties (Whitham, 1974; Miles, 1981), and many aspects of the breaking process in shallow water are revealed by the study of solitary waves of high amplitude, *i.e.* close to the limiting ϵ_w (Longuet-Higgins & Fenton, 1974).

The well-known first order solitary wave theory can be obtained as the limiting solution of the KdV equation with infinite wavelength, yet having a finite effective length. In a reference frame moving with the wave celerity (the wave crest being located at $x=0$), the solution is

$$\frac{\eta}{D}=1+\epsilon_w sech^2(\alpha_w\frac{x}{D}) \tag{3.8}$$

$$-\frac{u_w}{(gD)^{1/2}}=1+\frac{1}{2}\epsilon_w-\epsilon_w sech^2(\alpha_w\frac{x}{D}) \tag{3.9}$$

$$\frac{v_w}{(gD)^{1/2}}=(3\epsilon_w)^{\frac{1}{2}}\epsilon_w\frac{y}{D}sech^2(\alpha_w\frac{x}{D})tanh(\alpha_w\frac{x}{D}) \tag{3.10}$$

where $\alpha_w=(3/4\epsilon_w)^{1/2}$ is a straining parameter (Lighthill, 1949). D/α_w is an effective length of the solitary wave. In terms of H, D and D/α_w the Ursell number for the solitary wave is expressed as

$$U_r = \frac{H}{D\alpha_w^2} = \frac{4}{3} \sim \mathcal{O}(1)$$

This shows the essential quality of the solitary wave: the perfect balance between nonlinear steepening and dispersive effects. The wave celerity c (dispersion relation for the solitary wave) is

$$\frac{c}{(gD)^{1/2}} = \mathbf{F} = 1 + \frac{1}{2}\frac{H}{D} = 1 + \frac{1}{2}\epsilon_w$$

where $\mathbf{F}$ is the Froude number (the solitary wave is supercritical). Unidirectional first-order solitary waves of different amplitudes pass through one another without loss of identity and suffer only phase shifts (see *e.g.* Whitham (1974) for an extensive discussion of this topic), and thus are solitons. Higher order solutions may <u>not</u> be solitons.

Using a very high order expansion, a especially developed expansion parameter and convergence improvement techniques, Longuet-Higgins & Fenton (1974) demonstrated that the wave celerity, mass M, momentum I and total energy E

$$M = \int_{-\infty}^{+\infty} (\eta - 1)\, dx$$

$$I = \int_{-\infty}^{+\infty}\int_0^{\eta} u_w\, dx dy$$

$$E = \int_{-\infty}^{+\infty}\int_0^{\eta} \frac{1}{2}(u_w^2 + v_w^2)\, dx dy + \int_{-\infty}^{+\infty} \frac{1}{2} g (\eta - 1)^2\, dx$$

attain local maxima at a certain ϵ_w less than the maximum, which was found to be 0.827. Figure 3.7 (after Longuet-Higgins & Fenton, 1974) shows the dimensionless momentum, mass and total energy as functions of ϵ_w (all quantities are made nondimensional by choosing time and length scales such that $D = 1$ and $g = 1$). The properties of high solitary waves explain the qualitative difference between plunging and spilling breakers. The total energy $E(gD^3)$ will remain approximately constant for a solitary wave entering a mild-sloped bottom. Hence, both ϵ_w and E will increase at first. When the maximum E is attained the wave cannot remain symmetrical and steady, and may evolve towards the symmetrical wave of maximum amplitude. Instabilities in the wave front then induce small breaking and dissipation.

If the variation of D is slow the wave may jump again to a new ϵ_w below the maximum of E. In this way the wave may grow up again and the cycle breaking-nonbreaking is restarted, explaining the intermittence that has been observed in spilling breakers. If the bottom slope is steep, the wave arrives at the maximum of E without dissipating enough energy, deforms rapidly, and becomes a plunging breaker (Longuet-Higgins & Fenton, 1974). This complex behavior of solitary waves of high amplitude cannot be studied in terms of the Boussinesq/KdV theory.

In the present work, solitary wave theories are used to generate initial and boundary conditions outside the breaker area. Fenton's (1972) third-order expressions for the solitary wave fluid velocity, surface profile and pressure were used, because the main interest was concentrated on highly nonlinear waves.

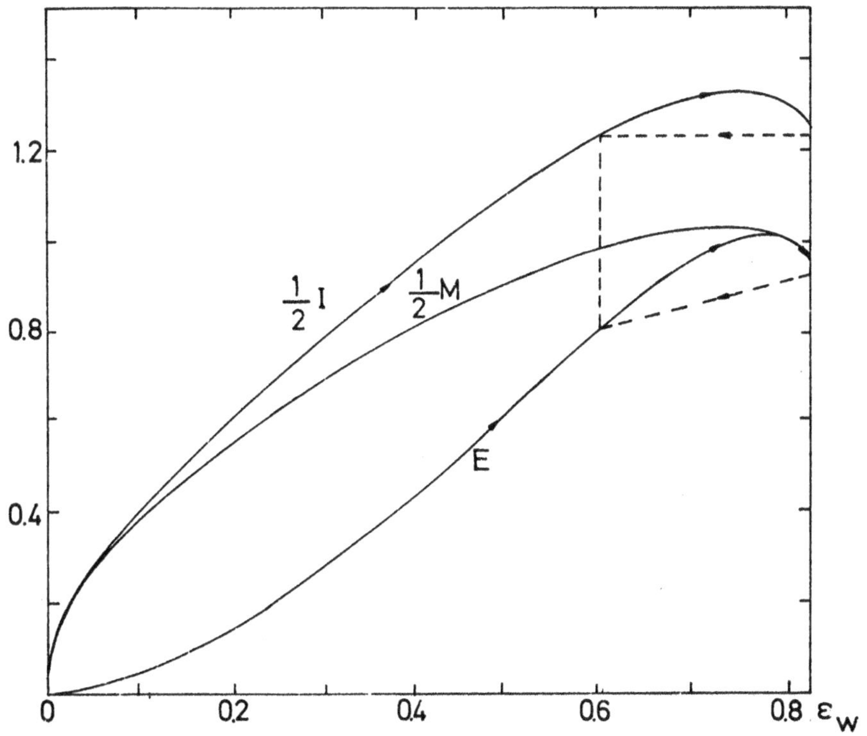

Figure 3.7 : *Dimensionless momentum I, mass M and total energy E as functions of the dimensionless wave height (after Longuet-Higgins & Fenton,* 1974).

3.4.2 Hydraulic jumps. Discrete forms of the conservation laws.

A hydraulic jump is a sudden transition to subcritical flow that may occur in a supercritical stream, characterized by an increase in water level and a loss of energy. Figure 3.8 schematizes a stationary hydraulic jump, which is ideally treated as a step discontinuity. The usual assumptions of classical hydraulic jump theory are: the jump is stationary; the bottom is horizontal; the horizontal velocity is uniform over the entire depth at sections $\mathcal{S}_1$ and $\mathcal{S}_2$; the vertical acceleration is negligible at $\mathcal{S}_1$ and $\mathcal{S}_2$ (*i.e.* pressure is hydrostatic); and bottom friction effects are small. Both $\mathcal{S}_1$ and $\mathcal{S}_2$ are located in the regions of nearly-horizontal flow. The supercritical Froude number

$$\mathbf{F} = \frac{u_1}{(gh_1)^{1/2}}$$

determines whether a jump is undular, transitional, or turbulent, the geometric configuration, and many other jump properties (Rajaratnam, 1967).

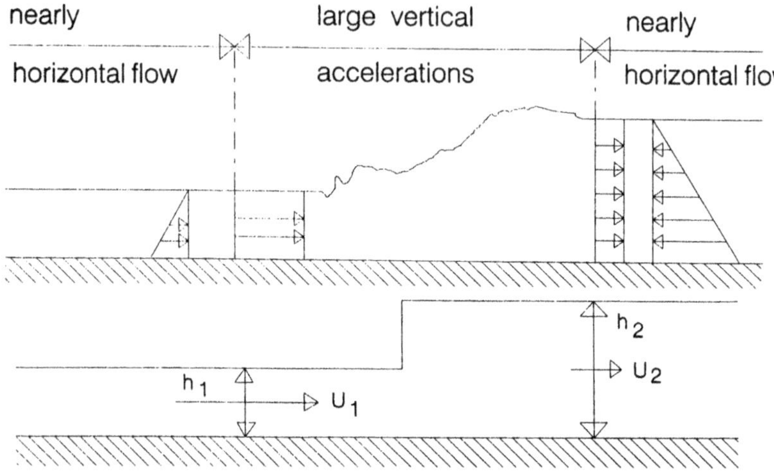

Figure 3.8 : *Schematization of the standing hydraulic jump.*

In § 3.3.1 it was seen that Q, R and S are invariant for a wave system without friction or other losses. This is no longer possible in a steady transition between two states of nearly horizontal flow. Conservation of mass and momentum (*i.e.* Q and S) between $\mathcal{J}_1$ and $\mathcal{J}_2$ implies (Lamb, 1932)

$$u_1h_1 = u_2h_2 = [\frac{1}{2}gh_1h_2(h_1+h_2)]^{1/2} \tag{3.11}$$

where $h_2 > h_1$. However (3.11) implies that $R_2 = 1/2u_2^2 + gh_2 < R_1 = 1/2u_1^2 + gh_1^2$, and thus energy is lost by the nearly-horizontal flow across the discontinuity. The rate of energy dissipation per unit span when passing from $\mathcal{J}_1$ to $\mathcal{J}_2$ is

$$\Delta E = \frac{1}{4}g\frac{(h_2-h_1)^3}{h_1h_2}\rho Q > 0 \tag{3.12}$$

According to the first principle of thermodynamics, the lost energy must be transferred to other types of motion, not of the nearly-horizontal flow type. According to the second principle of thermodynamics, these motions must be less organized than a nearly-horizontal flow, and hence may be either short-wave trains (dispersive loss) or turbulence (frictional loss) (Abbott, 1979). The hydraulic jump equations may also be obtained from the shallow-water equations via the weak solution concept. However, the weak solution approach does not explain the existence of undular jumps. Madsen & Svendsen (1983) and Svendsen & Madsen (1984) formulated extensions of the hydraulic jump allowing a moderate representation of the velocity and turbulence structure in the jump. The momentum and energy defect equations may be expressed in dimensionless form in terms of $\mathbb{F}$

$$\frac{h_2}{h_1}=\frac{1}{2}(\sqrt{1+8\mathbb{F}^2}-1) \qquad \text{(Bélanger, 1828)} \qquad (3.13)$$

$$\frac{\Delta E}{E_1}=\frac{1}{8}\frac{(\sqrt{1+8\mathbb{F}^2}-3)^3}{(2+\mathbb{F}^2)(\sqrt{1+8\mathbb{F}^2}-1)} \qquad \text{(Rajaratnam, 1967)} \qquad (3.14)$$

where E_1 is the incident energy flux per unit span. Favre (1935) found experimentally that when $h_2/h_1 < 1.28$ ($\mathbb{F} < 1.21$) the jump develops a stationary wave train behind it (undular jump), whereas when $h_2/h_1 > 1.28$ the jump breaks, inducing strong turbulence.

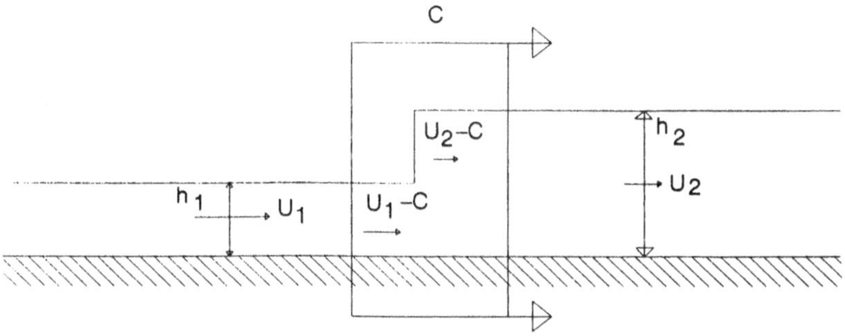

Figure 3.9 : *Relationships between velocities in a reference frame moving with a hydraulic jump, and a fixed reference frame.*

Most problems involve moving rather than steady jumps. If the effect of bottom dissipation is neglected, the equations of conservation of mass and momentum are identical for all Galilean frames (figure 3.9). For a jump traveling with a celerity C, the mass and momentum conservation equations are

$$C[h]_1^2=[uh]_1^2$$

$$C[uh]_1^2=[u^2h+\frac{1}{2}gh^2]_1^2$$

where $[\ ]_1^2$ denotes the jump of the quantity within brackets across the discontinuity.

Because breaking waves in the inner region resemble moving hydraulic jumps, or bores, the hydraulic jump plays an important role in the present work. The computed h_2/h_1 and C for moving hydraulic jumps, in both undular and breaking regimes, provides a precise check on the mass and momentum conservation in the numerical model. Also, it is important to determine the structure of the flow in both regimes, the nature of the transition at the critical h_2/h_1 , the geometric characteristics of undular and breaking bores, and the ability of the K-ϵ model for representing the structure of the turbulence field inside the body of turbulent jumps.

3.5 Summary of experimental investigations.

In this section, a summary of experimental investigations that are relevant for comparison with the numerical results obtained in this work is presented.

3.5.1 Wave-wave interactions.

Wave-wave interaction is a complicated phenomenon, which has a considerable importance in wave breaking (Benjamin & Feir, 1967; Longuet-Higgins, 1978a, 1978b). This phenomenon has been studied for both periodic and solitary waves using analytical, numerical, and experimental methods. Interaction between deep and shallow water waves is qualitatively different, because in deep water waves the influences propagate in a half-plane of fluid whereas in shallow water waves the disturbances propagate in a more restricted domain at speeds that cannot be much greater than $(gD)^{1/2}$ (Peregrine, 1988). In this work, the interest is in very shallow water waves, and consequently the collisions between solitary waves are especially relevant. This type of wave-wave interaction has received great attention. Although soliton theory is well developed and extensive, experimental investigations show that for highly nonlinear waves the theoretical predictions are not always verified. The interaction problem is completely specified by the dimensionless wave heights of the colliding waves. There are two quite distinct types of interactions: overtaking, and head-on collisions.

To first order in wave height and steepness, the collision of overtaking solitary waves is described by the KdV equation. An exact solution for N solitary waves was found (see *e.g.* Whitham, 1974 and Miles, 1980) which shows that overtaking first-order solitary waves are solitons. Weidman & Maxworthy (1978) performed experiments on this type of interaction and found good agreement with the KdV theory.

In the case of head-on collision between solitary waves, the interaction is brief but much more nonlinear. The theory of such collisions is less well developed, but it is known that high-order effects are very important (Schwartz & Fenton, 1982). Fenton & Rienecker (1982) performed numerical calculations of head-on collisions, and showed that the waves became steeper after the interaction, their form being degraded.

Maxworthy (1976) investigated experimentally the head-on collision of two solitary waves of equal amplitude. This type of collision is interesting because it depends on a single parameter (the dimensionless wave height of the colliding waves), and because it is physically identical to the reflection of a single wave by a vertical free-slip wall (located at the abscissa of the center of mass of the two waves). The experimental investigation consisted in the determination of the wave speed and profile, the phase shift and the maximum wave height, for both collisions between two equal waves and reflections at a wall. Dimensionless amplitudes of 0.1 to 0.5 were used. It was found (Maxworthy, 1976) that the wave speed before the interaction and the maximum fluid height during the collision was in good agreement with first-order theory based on the Boussinesq equations, but the observed phase shift was not. Theory predicted a phase shift increasing with ϵ_w, but the observed phase shift was nearly constant for all ϵ_w.

Head-on collisions between solitary waves are important in the context of the present work (see §§ 6.3.1 and 6.3.2). This phenomenon is a simplified model for the interaction between incident and reflected waves in the outer region of the surf-zone. The numerical approach here developed is potentially able to analyze the detailed structure of the flow before, during, and after the interaction, and the resulting development of instabilities or deformations.

3.5.2 Hydraulic jumps.

The relevant investigations on the hydraulic jump are of two types: those concerned with undular and transitional jumps, and those concerned with turbulent jumps. This ordering is suggested by the practical interest in describing how the jump changes from dispersive to dissipative regime. It is also an ordering based on $\mathbf{F}$, which is the relevant dimensionless parameter.

Favre (1935) performed a number of experiments on the propagation of a hydraulic jump, for several sequent depth ratios. His main conclusions are :

- For $h_2/h_1 < 1.28$ ($\mathbf{F} < 1.21$), the propagating jump maintains a purely undular character. For $h_2/h_1 > 1.28$ breaking becomes increasingly prominent, inducing intense turbulence within the body of the jump.

- The number of waves present at any instant in undular jumps is proportional to the time elapsing from the initiation of the jump.

- The waves radiated by undular jumps are remarkably uniform and have a regular period. A unique amplitude appears for each pair of sequent depths and time T.

- The wavelengths are exceptionally sensitive to variations in incident energy or momentum fluxes, hence to experimental conditions.

The first conclusion suggests that $\Delta h/h_1$ is a basic instability parameter with an associated critical value. The second conclusion supports the viewpoint that the wave group carries energy away from the jump. Lemoine (1948) found approximate expressions for the amplitude and wavelength of jump waves

$$\frac{a}{h} \sim \frac{1}{\sqrt{3}}\frac{h_2-h_1}{h_1} \tag{3.15}$$

$$\frac{\lambda}{h_2} \sim \frac{2\pi\sqrt{2}}{3}\left(\frac{h_2-h_1}{h_1}\right)^{-\frac{1}{2}} \tag{3.16}$$

For $h_2/h_1 = 1.28$, $a/h_2 \sim 0.16$. The amplitude is large because of the small difference between the phase and group celerities (Benjamin & Lighthill, 1954), which is the celerity at which the energy is radiated. Equation (3.16) shows that λ/h_2 is sensitive to Δh, in particular for very weak jumps. Abbott & Rodenhuis (1972) performed numerical simulations of Favre waves, using a Boussinesq-type system of equations. They found that the computed wavelengths were very sensitive to the truncation errors, and that the computed critical ratio h_2/h_1 was 1.19 instead of 1.28. Peregrine (1974) pointed out that the equation system used by these authors is not formally consistent. Abbott & Rodenhuis (1974) replied to Peregrine's critic, and defended that for an ideal fluid all undular jumps will ultimately break.

For $\mathbf{F} \geq 2$, the jump is fully turbulent ("strong"). A roller is formed in the region where the water level suddenly rises. Seen from an appropriate Galilean frame, the jump is then similar to a fully developed (inner region) breaker. Hydraulic jumps in such conditions have been object of several investigations. These studies were mainly performed on stationary jumps, and include verification of the momentum (h_2/h_1) equation, description of the surface profile, mean velocity and turbulence characteristics, air entrainment characteristics, and the effect of bottom shear stress in the momentum equation. A detailed survey of experimental investigations up to 1967 may be found in Rajaratnam (1967). More recent data and discussions on aspects of the turbulence field may be found in Resh & Leutheusser (1972), and in Resh et al. (1976). When scaled by the appropriate length or velocity, the geometric and dynamic properties are functions of $\mathbf{F}$. Figure 3.10 (after Resh et al. , 1976) shows typical mean velocity streamlines for two hydraulic jumps with $\mathbf{F} = 2.85$ for different inflow conditions. Figure 3.11 (after Resh & Leutheusser, 1972) represents the dimensionless intensity of the velocity fluctuations in the x and y directions, and the Reynolds stresses in a turbulent hydraulic jump.

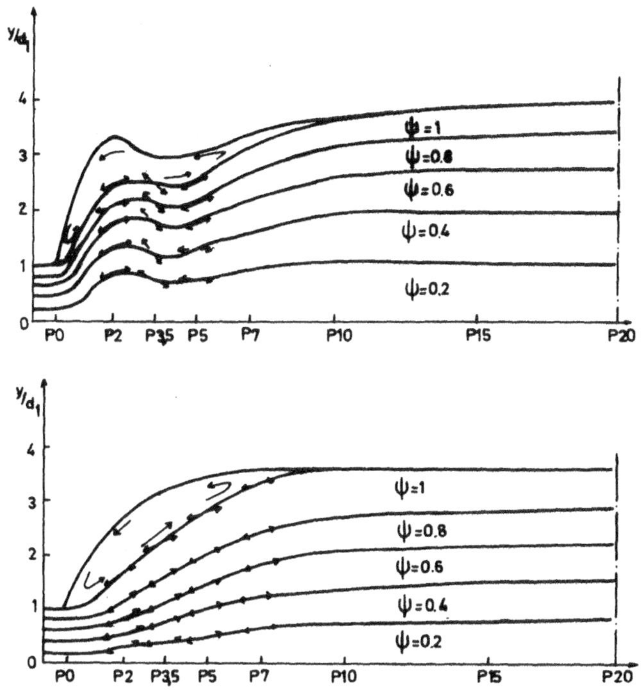

Figure 3.10 : *Mean streamline patterns in turbulent hydraulic jumps,* **F** = 2.85. *Top, fully developed inflow; bottom, potential inflow* (*after Resh et al.* , 1976).

From the above discussion it is concluded that the hydraulic jump is an important type of shallow water flow. Previous studies leave some important questions open, which are important for understanding the breaking phenomenon. First, the correct simulation of Favre waves and the critical limit for transition is an important test for the present approach (§ 6.4.1). For example, it would be interesting to determine if the flow in a hydraulic jump can be interpreted in terms of the sequence *disturbances* ⇒ *evolution* ⇒ *breaking*, which is essentially the same used for other types of waves (§ 6.4.2). Second, the numerical simulation of the mean flow and turbulence field properties in a strong (fully turbulent) jump is an important challenge (§ 6.4.3). State-of-the-art models are not yet able to solve this problem with sufficient generality.

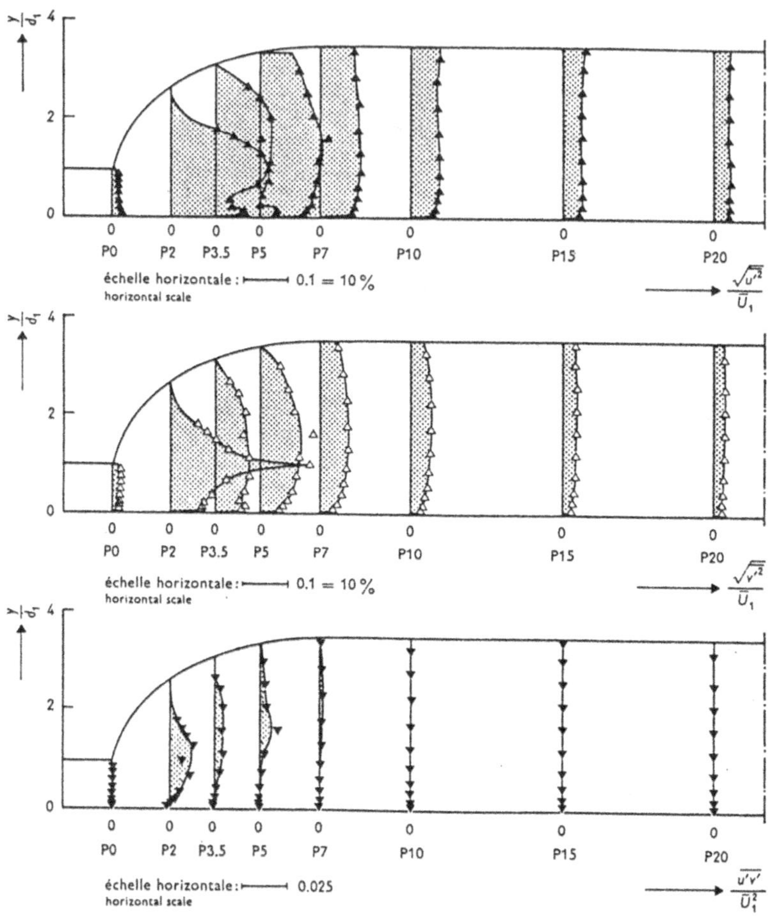

Figure 3.11 : *Intensity of velocity fluctuations in a hydraulic jump (after Resh & Leutheusser*, 1972).

3.5.3 Waves breaking on a slope.

Recently, a number of experimental data on the detailed structure of the velocity and turbulence fields in breaking waves has been published (*e.g.* Mizuguchi, 1986 , Nadaoka & Kondoh, 1982; Nadaoka, 1986). These data have been obtained in the laboratory, and allow a detailed qualitative and quantitative analysis of the flow field in broken waves in all stages of breaking.

The experimental study of Mizuguchi (1986) is a good representative of such works. This study was performed in a wave flume with a slope $m = 1/20$. Periodic waves with a period $T = 1.22\ s$ and a deep water height of $H_0 = 4.4 \times 10^{-2}\ m$ were used, giving a breaker of the plunging type with $H_{br} = 6.1 \times 10^{-2}\ m$. The velocity field was sampled by means of an array of Laser-Doppler sensors. Phase averaging (which is a particular case of ensemble averaging) was used to separate the wave and turbulent components. The field variables $\vec{U}_w$, $\nabla\cdot\vec{U}_w$. $\nabla\times\vec{U}_w$. $\vec{A}_w = D\vec{U}_w/Dt$. which are the mean velocity, divergence, curl and particle acceleration, and the turbulent kinetic energy $K = 1/2\overline{u_i u_i}$ were obtained in several stages of the breaking process. Figure 3.18 represents the velocity, divergence, vorticity and turbulent kinetic energy for a wave just after the plunging has started.

The overturning motion is seen to be nearly irrotational. The velocity structure resembles that of a solitary wave, but deformed by the wave evolution due to the large bottom slope (§ 3.4.1). There is residual turbulence in the front of the wave, left by the previous wave. Production of vorticity and turbulence from the breaking wave starts only after plunging. The degree of turbulence isotropy is indicated by the slope of the small segments in figure 3.12 . It is seen that turbulence is nearly isotropic in most regions of the flow, except near the bottom due to the effect of the solid boundary. This suggests that the main features of the surface-induced turbulence may be reasonably well described using an eddy-viscosity model.

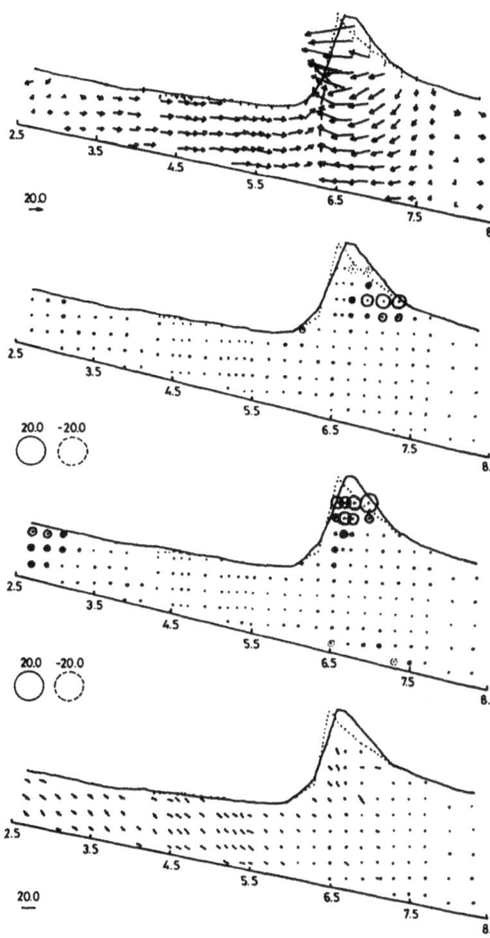

Figure 3.12 : *Velocity, divergence, vorticity and turbulence intensity in a broken wave* (*after Mizuguchi*, 1986).

A later stage of the breaker is depicted in figure 3.13 . It is seen that vorticity and turbulence are being created close to the surface. Production and dissipation of turbulence and vorticity seem to be more important than convection and diffusion, because both are dissipated near to the place where they were created. This is also supported by Svendsen's (1987) investigation of surf-zone turbulence. The turbulence intensity persists longer than vorticity. The wave form and internal structure approach the bore model as the wave advances into shallower water.

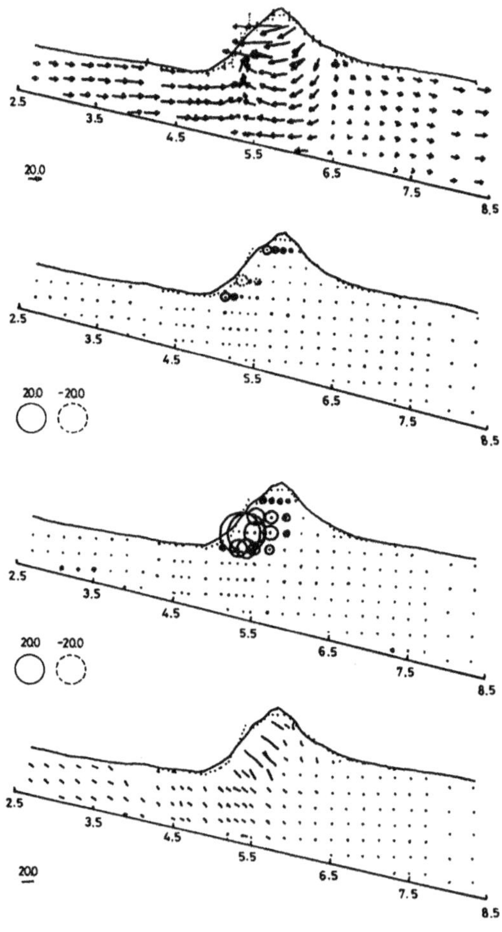

Figure 3.13 : *Velocity, divergence, vorticity and turbulence intensity in a broken wave (continued)* (*after Mizuguchi*, 1986).

Other important aspects of the structure of breaking waves are revealed by the experimental study of Nadaoka (1986). The experimental equipment used is described in figure 3.14. The experiment was performed using waves with period T=1.36 s. The depth and wave height of the breaker were 14.7×10^{-2} m and 14.9×10^{-2} m, respectively. Separation of the turbulent component was performed by both phase average and filtering methods (Nadaoka, 1986). The filtered distribution of the horizontal and vertical components of the turbulent fluctuation are represented in figure 3.15.

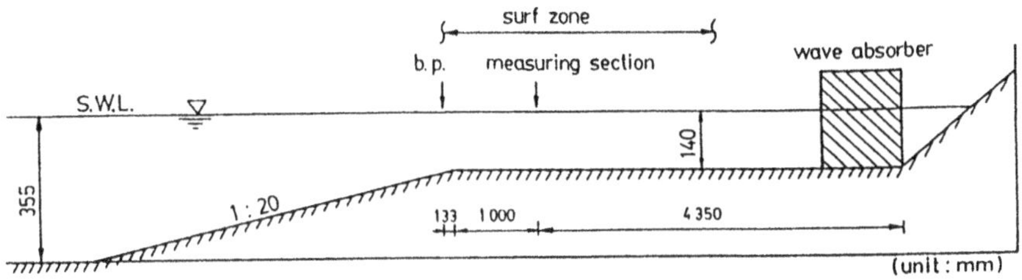

Figure 3.14 : *Experimental equipment used in Nadaoka's study of breaking waves (after Nadaoka, 1986).*

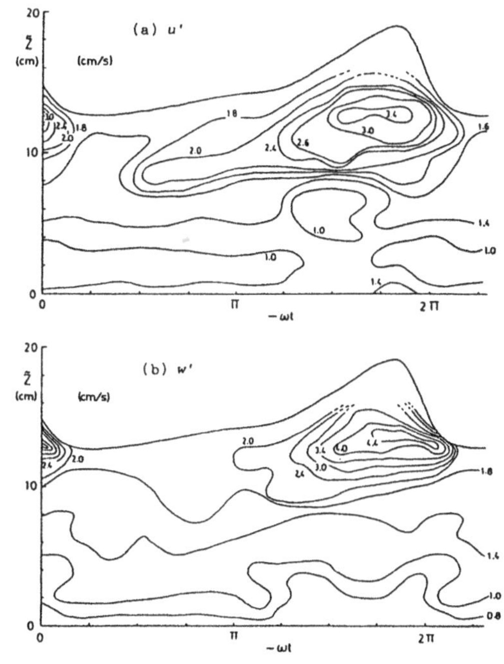

Figure 3.15 : *Distribution of the horizontal and vertical components of the velocity fluctuation, detected by a moving average method (after Nadaoka, 1986).*

The Reynolds stress detected by a moving average method is represented in figure 3.16. The Reynolds stress detected by a phase average method is shown in figure 3.17. In both cases, the Reynolds stress attains its peak value at the lower edge of the eddy region under the wave trough. The magnitude of the phase average method is two orders of magnitude larger than that by the moving average (filtering) method. This shows the difficulty of separating turbulence from the other types of motion in wave breaking experiments.

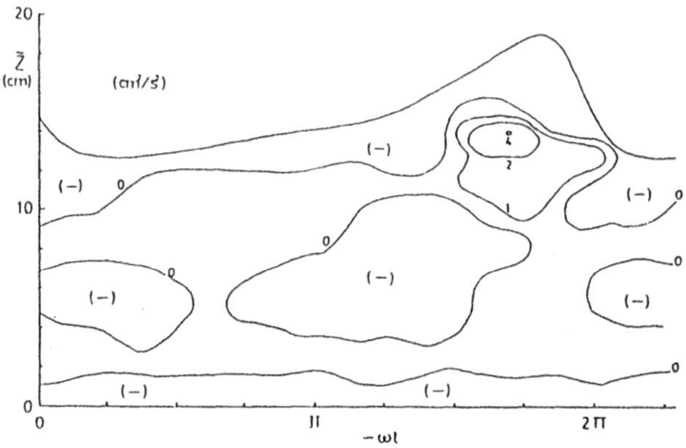

Figure 3.16 : *Reynolds stress detected by a moving average method (after Nadaoka, 1986).*

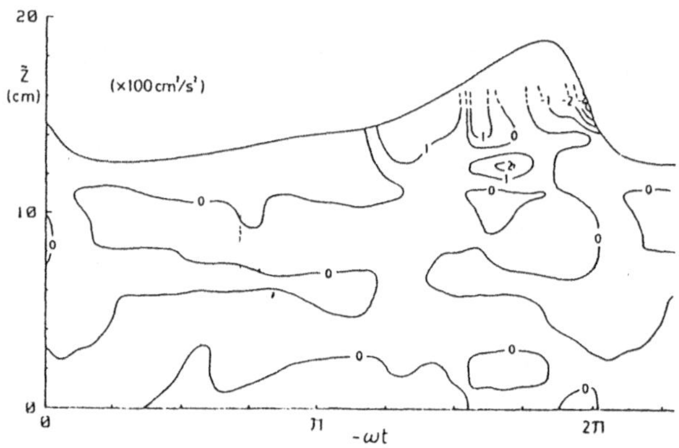

Figure 3.17 : *Reynolds stress detected by a phase average method (after Nadaoka, 1986).*

The experiments of Nadaoka (1986) and Mizuguchi (1986) provide a useful orientation for the setup and interpretation of the numerical solutions obtained in the present work (§ 6.5). First, the type and quality of the data are adequate for the present numerical approach. Second, the space and time scales of the experiment allow a numerical simulation with a good spatial resolution and realistic molecular and turbulent viscosity coefficients. However, in real surf-zones, the physical dimensions of the domain and the length scales for the wave motion are one to two orders of magnitude larger. Therefore, the significance of these laboratory experiments in field situations is a matter of discussion. Third, the relatively steep bottom slope produces a clear plunging breaker, which evolves rapidly (requiring less computer time to reach significant conclusions) and is more consequent with the simplifying hypotheses in the present model, such as incompressibility and constant density.

3.6 Description of the proposed methodology.

The physical processes involved in the transformation of normally-incident waves on a beach are very complicated. A general mathematical representation of these physical processes is not available. Several models exist that are capable of a partial representation of these phenomena, but none can describe all of them simultaneously. This situation is summarized in table 3.2, by showing the ability of the available mathematical methods for representing breaking-wave physical processes.

Table 3.2 : Wave transformation—Physical processes vs. mathematical descriptions.

Type of model	Physical process						
	Nonlinear steepening	Dispersion	Wave deformation	Multiple free-surf.	Vorticity	Turbulent transport	Air-water mixing
Wave theories	*Yes*	*Yes*	*No*	*No*	*No*	*No*	*No*
SWE's	*Yes*	*No*	*Yes*	*No*	*No*	*No*	*No*
Boussinesq SWE's	*Yes*	*Yes*	*Yes*	*No*	*No*	*No*	*No*
Time-dependent potential flow models	*Yes*	*Yes*	*Yes*	*No*	*No*	*No*	*No*
MAC-type models	*Yes*	*Yes*	*Yes*	*Yes*	*Yes*	*No*	*No*

MAC-type methods emerge from the above comparison as the most promising approach.

They have less *a-priori* restrictions, and can represent overturning, arbitrary free-surface configurations, and vorticity, and hence are more compatible with the complexity of the physical processes than the other mathematical methods. MAC-type models are well established, and have been successfully used in solving difficult free-surface flow problems. Because the full momentum equation is used, these models can be generalized to include turbulent diffusion, described by a suitable turbulence model. This methodology has been followed in other branches of computational fluid dynamics, but so far has not been proposed for wave breaking problems with the generality here considered.

In the present work, a MAC-type model for solving the time-dependent, two-dimensional Reynolds equations is coupled with an efficient numerical technique for tracking arbitrary free-surfaces (introduced by Nichols & Hirt, 1975) and a especially developed K-ϵ model. It will be shown that this formulation has a considerable potential for studying wave transformation phenomena. Many types of confined flows of interest, either laminar or turbulent, can also be investigated.

The proposed methodology will be developed in the following steps :

- A detailed revision of MAC-type methods for free-surface incompressible flow is performed, with a special emphasis on the techniques for pressure solution and free-surface treatment. Then, the basic elements of an efficient algorithm are selected.

- The incorporation of the Reynolds stress terms in the solution algorithm is investigated, as well as the general principles for discretization of the K-ϵ equations.

- A computer program, based on the solution algorithm for free-surface incompressible flows of arbitrary configuration and a especially developed K-ϵ turbulence model is constructed. A preliminary validation is performed, by comparing the computed results with theoretical predictions for various test problems concerning laminar and turbulent flows, either confined or involving free-surfaces.

- The computer program is applied to selected problems for shallow-water wave transformation, concerning wave propagation and interaction prior to breaking, transition to breaking regime, and the propagation of broken waves.

Chapter 4

MAC—TYPE METHODS FOR INCOMPRESSIBLE FREE—SURFACE FLOWS

4.1 Introduction.

In this chapter, a revision of MAC (Marker-and-cell) type methods is presented. These methods consist of a solution algorithm for the incompressible NS equations in the primitive-variable approach, and a technique for defining and updating the free-surface. In MAC-type methods, the momentum equations are typically discretized by means of 5-point (in two dimensions) forward-Euler methods, which greatly simplify the application of several types of boundary conditions, the free-surface treatment, and the relationship between the velocity and pressure solutions. MAC-type methods have been the basis for solving many difficult free-surface flow problems (Welch et al. , 1966; Amsden & Harlow, 1970; Nichols et al. , 1980; Miyata, 1986), and hence are well established. They are also well adapted for incorporation of turbulence models and implementation in vector and parallel computers. The efficiency of each particular variant of the basic MAC algorithm is strongly dependent on two technical problems: the pressure solution and the method for defining and updating the free-surface.

The outline of this chapter is as follows. In § 4.2 , the arrangement of the dependent variables in the discrete mesh is discussed. In §§ 4.3—4.6 , a comparative review of finite-difference methods is presented. § 4.7 deals with the problem of representing a free-surface in a discrete model. In § 4.8, some stability results of general applicability are reviewed. Finally, the characteristics of the solution algorithm employed in this work are presented.

4.2 The choice of the mesh.

Before discretizing the governing equations, the mesh to be used must be selected. In this work, rectangular meshes of variable size are used. This is not as flexible as using curvilinear meshes, but allows a certain degree of boundary resolution (which is particularly important for the correct application of the law-of-the-wall in the turbulence model).

Finite-difference incompressible flow calculations are mainly performed on two types of mesh arrangements: the staggered (MAC) mesh and the centered mesh. The location of the dependent variables in each case is represented in figure 4.1. In this work, the MAC mesh is used exclusively.

The centered mesh seems the simplest choice, but has some disadvantages with respect to the MAC mesh: pressure must be defined at the boundaries and the pressure solution is prone to even-odd decoupling. In the MAC mesh, pressure is not defined at the boundaries, and there exists a tight relationship between the pressure and velocity fields. This type of mesh does not originate decoupling problems.

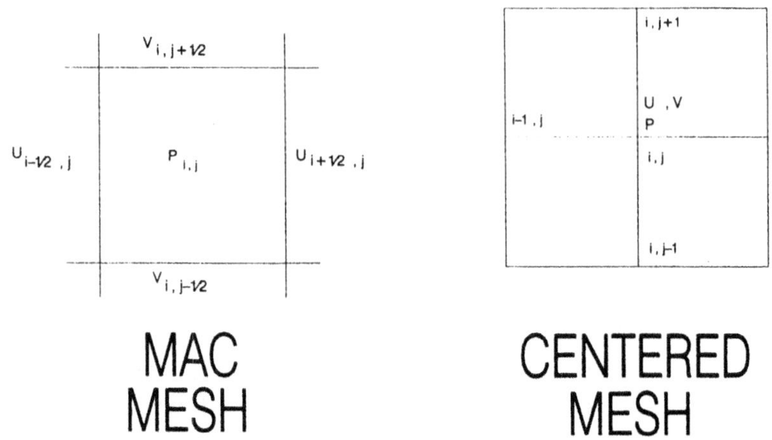

Figure 4.1 : *Location of dependent variables in the staggered (MAC) mesh and in the centered mesh*

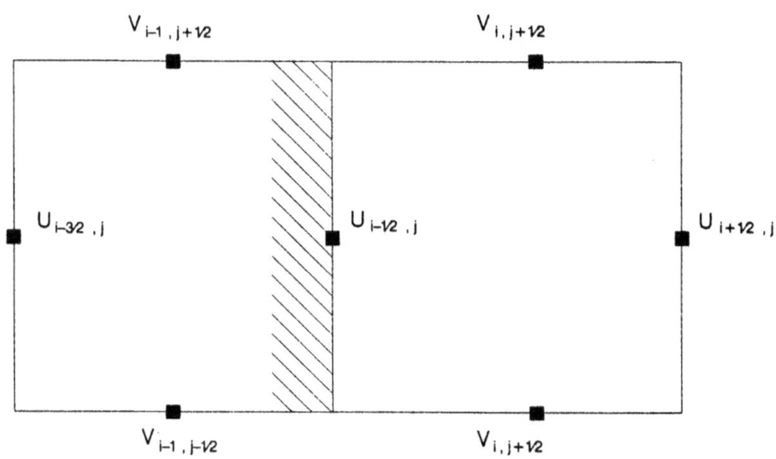

Figure 4.2 : *Velocity boundary conditions in the MAC mesh*

Though efficient in terms of pressure treatment, the MAC mesh presents a disadvantage concerning the velocities. Figure 4.2 represents the MAC mesh arrangement close to a wall boundary Σ, where a velocity (u_Σ, v_Σ) is to be prescribed as a boundary condition. The velocity component normal to the boundary (u_Σ) can be specified at the correct location by simply setting $u_{i-1/2,j} = u_\Sigma$. However, the tangential velocity (v_Σ) is not available at the correct place in the discrete mesh, *i.e.* $v_{i-1,j-1/2}$ and $v_{i-1,j+1/2}$ are not defined on Σ. In order to prescribe the tangential velocity on Σ, $v_{i-1,j-1/2}$ and $v_{i-1,j+1/2}$ must be related to v_Σ. The simplest method of doing this consists of defining $v_{i-1,j+1/2}$ ($v_{i-1,j-1/2}$) as a linear extrapolation between v_Σ and $v_{i,j+1/2}$ ($v_{i,j-1/2}$)

$$v_{i-1,j+1/2} = v_\Sigma + \frac{\Delta x_{i-1}}{\Delta x_i}(v_\Sigma - v_{i,j+1/2})$$

The "outside value" $v_{i-1,j+1/2}$ is then used to compute the approximations for $u(\partial v/\partial x)$ and $\nu(\partial^2 v/\partial x^2)$ in the momentum equations. When writing down finite difference equations in the MAC mesh, it is convenient to define a number of finite-difference operators. For a regular mesh with $\Delta x_i = \Delta x$, $\Delta y_j = \Delta y$ these operators are

$$\Delta_x^1 f_{i,j} = \frac{1}{\Delta x}(f_{i+1/2,j} - f_{i-1/2,j})$$

$$\Delta_y^1 f_{i,j} = \frac{1}{\Delta y}(f_{i,j+1/2} - f_{i,j-1/2})$$

$$\Delta_x^0 f_{i,j} = \frac{1}{2\Delta x}(f_{i+1,j} - f_{i-1,j}) \quad ; \quad \Delta_y^0 f_{i,j} = \frac{1}{2\Delta y}(f_{i,j+1} - f_{i,j-1})$$

$$\Delta_x^+ f_{i,j} = \frac{1}{\Delta x}(f_{i+1,j} - f_{i,j}) \quad ; \quad \Delta_x^- f_{i,j} = \frac{1}{\Delta x}(f_{i,j} - f_{i-1,j})$$

$$\Delta_y^+ f_{i,j} = \frac{1}{\Delta y}(f_{i,j+1} - f_{i,j}) \quad ; \quad \Delta_y^- f_{i,j} = \frac{1}{\Delta y}(f_{i,j} - f_{i,j-1})$$

$$\Delta_{xx} f_{i,j} = \Delta_x^1 \Delta_x^1 f_{i,j} = \frac{1}{\Delta x^2}(f_{i+1,j} - 2f_{i,j} + f_{i-1,j})$$

$$\Delta_{yy} f_{i,j} = \Delta_y^1 \Delta_y^1 f_{i,j} = \frac{1}{\Delta y^2}(f_{i,j+1} - 2f_{i,j} + f_{i,j-1})$$

$$\nabla_h^2 f_{i,j} = \Delta_{xx} f_{i,j} + \Delta_{yy} f_{i,j}$$

If the mesh is irregular (figure 4.3), the forward, backward and second-order central differences, and the second-order approximations for the second derivatives are defined by

$$\Delta_x^+ f_{i,j} = \frac{1}{\Delta x^+}(f_{i+1,j} - f_{i,j}) \quad ; \quad \Delta_x^- f_{i,j} = \frac{1}{\Delta x^-}(f_{i,j} - f_{i-1,j})$$

$$\Delta_y^+ f_{i,j} = \frac{1}{\Delta y^+}(f_{i,j+1} - f_{i,j}) \quad ; \quad \Delta_y^- f_{i,j} = \frac{1}{\Delta y^-}(f_{i,j} - f_{i,j-1})$$

$$\Delta_x^0 f_{i,j} = \frac{1}{\Delta x^+ + \Delta x^-}(\Delta x^-[\Delta_x^+ f_{i,j}] + \Delta x^+[\Delta_x^- f_{i,j}])$$

$$\Delta_y^0 f_{ij} = \frac{1}{\Delta y^+ + \Delta y^-}(\Delta y^-[\Delta_y^+ f_{i,j}] + \Delta y^+[\Delta_y^- f_{i,j}])$$

$$\Delta_{xx} f_{i,j} = \frac{2}{\Delta x^+ + \Delta x^-}(\Delta_x^+ f_{i,j} - \Delta_x^- f_{i,j})$$

$$\Delta_{yy} f_{i,j} = \frac{2}{\Delta y^+ + \Delta y^-}(\Delta_y^+ f_{i,j} - \Delta_y^- f_{i,j})$$

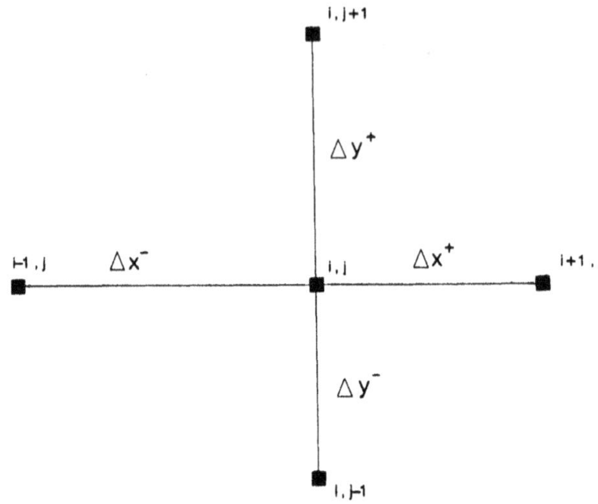

Figure 4.3 : *Notation for irregular meshes*

4.3 The MAC (Marker-and-Cell) method.

The MAC method (Welch et al. , 1966) was developed for the solution of viscous, incompressible and transient fluid flow problems involving free-surfaces. The primitive variables, pressure and velocity, are used directly as dependent variables. The fluid configuration is determined by means of a set of marker particles (massless tracers) moving with the fluid. These marker particles do not influence the dynamics, but define which computational cells are full of fluid, empty, or contain a free-surface. The MAC algorithm consists of the following steps :

1) The pressure in each cell is obtained by solving a finite-difference Poisson equation.

2) The momentum equations in the x and y directions are used to find the new velocities in all mesh cells, using the computed pressure field.

3) Marker particles are moved to their new positions, using interpolated values from nearly cells.

4) Mesh cells are reflagged as "full", "empty" or "surface", according to the new fluid configuration and the cycle restarted.

The MAC method is applied to the two-dimensional NS equations in conservative form

$$\frac{\partial u}{\partial t}+\frac{\partial}{\partial x}(u^2)+\frac{\partial}{\partial y}(uv) = -\frac{\partial \varphi}{\partial x}+\nu\left(\frac{\partial^2 u}{\partial x^2}+\frac{\partial^2 u}{\partial y^2}\right)+g_x \tag{4.1}$$

$$\frac{\partial v}{\partial t}+\frac{\partial}{\partial x}(uv)+\frac{\partial}{\partial y}(v^2) = -\frac{\partial \varphi}{\partial y}+\nu\left(\frac{\partial^2 v}{\partial x^2}+\frac{\partial^2 v}{\partial y^2}\right)+g_y \tag{4.2}$$

$$\frac{\partial u}{\partial x}+\frac{\partial v}{\partial y}=0 \tag{4.3}$$

where $\varphi \equiv$ ratio of pressure to (constant) density, which represents the contribution to the specific energy of the flow due to the pressure field. The discrete divergence in cell ij is

$$D_{i,j} \equiv \frac{1}{\Delta x}(u_{i+1/2,j}-u_{i-1/2,j})+\frac{1}{\Delta y}(v_{i,j+1/2}-v_{i,j-1/2}) \tag{4.4}$$

and the incompressibility condition requires $D_{i,j}=0$ for every cell at every time step. Equations (4.1) and (4.2) are discretized as follows

$$\frac{1}{\Delta t}(u^{n+1}_{i+1/2,j}-u^{n}_{i+1/2,j}) = -(F_x)^n_{i+1/2,j}-\Delta^1_x\varphi_{i+1/2,j}+\nu\nabla^2_h u^n_{i+1/2,j}+g_x \tag{4.5}$$

$$\frac{1}{\Delta t}(v^{n+1}_{i,j+1/2}-v^{n}_{i,j+1/2}) = -(F_y)^n_{i,j+1/2}-\Delta^1_y\varphi_{i,j+1/2}+\nu\nabla^2_h v^n_{i,j+1/2}+g_y \tag{4.6}$$

with

$$(F_x)^n_{i+1/2,j} = \Delta^1_x(u^2)^n_{i+1/2,j}+\Delta^1_y(uv)^n_{i+1/2,j}$$

$$(F_y)^n_{i,j+1/2} = \Delta^1_x(uv)^n_{i,j+1/2}+\Delta^1_y(v^2)^n_{i,j+1/2}$$

Before equations (4.5) and (4.6) are used to find u^{n+1} and v^{n+1}, φ is determined by a discrete Poisson equation. This equation is derived from the *discretized* momentum equations. Applying Δ^1_x to the discrete x-momentum equation, Δ^1_y to the discrete y-momentum equation and adding the results, the discrete PPE is found to be (Welch et al. , 1966)

$$\nabla^2_h\varphi_{i,j} = -Q^n_{i,j}-\frac{D^n_{i,j}}{\Delta t}-\nu\nabla^2_h D^n_{i,j} \tag{4.7}$$

with $Q^n_{i,j}=\{\Delta_{xx}(u^2)^n_{i,j}+\Delta_{yy}(v^2)^n_{i,j}+2\Delta^1_x\Delta^1_y(uv)^n_{i,j}\}$. The discrete PPE (4.7) is solved in the MAC method by a Newton-Raphson relaxation technique. Boundary conditions for pressure are obtained by projecting the discrete momentum equations in the direction normal to the wall. After all variables have been advanced in time, marker particles are moved and the fluid configuration is updated.

4.4 The projection method.

The interest of the projection method (Chorin, 1968) is the splitting between pressure and velocity solutions which leads to a relatively simple PPE. As a result, the method is much more efficient than the MAC method. In the projection method, the first step consists of finding a provisional value of the velocity field using the momentum equations *without* the pressure gradient. This provisional velocity field is called $\tilde{u}$, $\tilde{v}$ (tilde velocities). The discrete equations for $\tilde{u}$ and $\tilde{v}$ are

$$\frac{1}{\Delta t}(\tilde{u}_{i+1/2,j}-u^n_{i+1/2,j}) = -(F_x)^n_{i+1/2,j}+\nu\nabla^2_h u^n_{i+1/2,j}+g_x \tag{4.8}$$

$$\frac{1}{\Delta t}(\tilde{v}_{i,j+1/2}-v^n_{i,j+1/2}) = -(F_y)^n_{i,j+1/2}+\nu\nabla^2_h v^n_{i,j+1/2}+g_y \tag{4.9}$$

These equations, together with the boundary conditions for velocity, assure that the $\tilde{u}$, $\tilde{v}$ field carries the correct vorticity. In general, however, this temporary field will not satisfy the constraint $\nabla\cdot\vec{u}=0$. The second step consists of correcting the temporary velocity field so that the flow will still carry the correct vorticity, satisfy the incompressibility condition and hence be correct and uniquely determined by Helmholtz's decomposition theorem (Chorin & Marsden, 1979; Gustafson, 1987). The equations for the second step are

$$\frac{u^{n+1}_{i+1/2,j}-\tilde{u}_{i+1/2,j}}{\Delta t}+\Delta^1_x\varphi^{n+1}_{i+1/2,j}=0 \tag{4.10}$$

$$\frac{v^{n+1}_{i,j+1/2}-\tilde{v}_{i,j+1/2}}{\Delta t}+\Delta^1_y\varphi^{n+1}_{i,j+1/2}=0 \tag{4.11}$$

$$\Delta^1_x u^{n+1}_{i,j}+\Delta^1_y v^{n+1}_{i,j}=0 \tag{4.12}$$

The discrete PPE is derived by applying Δ^1_x to (4.10) and Δ^1_y to (4.11), adding the results and making use of (4.12)

$$\nabla^2_h\varphi^{n+1}_{i,j}=\frac{1}{\Delta t}\tilde{D}_{i,j} \tag{4.13}$$

where $\tilde{D}_{i,j}$ is the divergence left in the velocity field by the first step. After pressure is found from (4.13), equations (4.10) and (4.11) give u^{n+1} and v^{n+1}, and the cycle is complete. The boundary condition for φ is obtained by projecting the vector equation

$$\frac{\vec{u}^{\,n+1}-\vec{\tilde{u}}}{\Delta t}+\nabla\varphi^{n+1}=0$$

on the outward unit normal $\vec{n}$ to the boundary, thus obtaining a Neumann condition

$$\left(\frac{\partial\varphi}{\partial n}\right)^{n+1}_{\Sigma}=-\frac{1}{\Delta t}(\vec{u}^{\,n+1}_{\Sigma}-\vec{\tilde{u}}_{\Sigma})\cdot\vec{n} \tag{4.14}$$

where $\vec{\tilde{u}}_{\Sigma}$ is the (not yet defined) value of $\vec{\tilde{u}}$ on Σ. The compatibility condition for this

Poisson-Neumann problem is always satisfied, because

$$\frac{1}{\Delta t}\int_{\Omega} \nabla\cdot\tilde{\vec{u}}\, dV = -\frac{1}{\Delta t}\int_{\Sigma} (\vec{u}-\tilde{\vec{u}})\cdot\vec{n}\, dS$$

and $\int_{\Sigma}\vec{u}\cdot\vec{n}\, dS = 0$. The Poisson equation for the pressure has a much simpler density function than the one used in the MAC method. The essential feature of the projection method is that the numerical solution is *completely* independent of $\tilde{\vec{u}}$ (Peyret & Taylor, 1983). This happens because 1) $\tilde{\vec{u}}$ at inner points is independent of $\tilde{\vec{u}}_{\Sigma}$ since it is calculated explicitly, and 2) because $\tilde{\vec{u}}_{\Sigma}$ appears simultaneously in the Poisson density and in the Neumann boundary condition for pressure in the PPE, and thus cancels identically. In particular, it is possible to choose $\tilde{\vec{u}}_{\Sigma} = \vec{u}_{\Sigma}^{n+1}$ and get a homogeneous Neumann condition for φ. This condition is purely numerical, and does not imply that the $\partial\varphi/\partial n = 0$ at the boundary. The projection method has been used with success in confined flow calculations up to very high Reynolds numbers (Gustafson & Halasi, 1986, 1987).

4.5 The SMAC (Simplified Marker-and-Cell) method.

Though successfully employed in many calculations of complex flows, the original MAC method is excessively complicated in several aspects. This is particularly true for the pressure treatment. In the SMAC method (Amsden & Harlow, 1970), the true pressure is never calculated. The SMAC algorithm consists of the following steps :

1) A tentative field of advanced-time velocities is calculated by using an arbitrary potential field $\theta_{i,j}$, which satisfies the correct normal stress boundary condition at the free-surface.

2) The tentative velocities are modified to their final values in such a way that the vorticity left by the first step is preserved and the incompressibility constraint satisfied. A potential function $\psi_{i,j}$ (called pseudo-pressure) is determined by the requirement that it should convert the velocity field to one that satisfies $\nabla\cdot\vec{u}^{n+1} = 0$. This pseudo-pressure only needs homogeneous (Dirichlet or Neumann) boundary conditions.

3) Fluid configuration is defined and updated by a set of marker particles, like in the MAC method.

Discrete equations for the intermediate velocity field $\tilde{u}$, $\tilde{v}$ are derived from the momentum equations using an arbitrary potential field $\theta_{i,j}$, with the only requirement that it satisfies the normal free-surface stress condition

$$\frac{1}{\Delta t}(\tilde{u}_{i+1/2,j}-u^n_{i+1/2,j}) = -(F_x)^n_{i+1/2,j}+VISX+\Delta^1_x\theta_{i+1/2,j}+g_x \tag{4.15}$$

$$\frac{1}{\Delta t}(\tilde{v}_{i,j+1/2}-v^n_{i,j+1/2}) = -(F_y)^n_{i,j+1/2}+VISY+\Delta^1_y\theta_{i,j+1/2}+g_y \tag{4.16}$$

and

$$(F_x)^n_{i+1/2,j} = \frac{u^n_{i+3/2,j}u^n_{i+1/2,j}-u^n_{i+1/2,j}u^n_{i-1/2,j}}{\Delta x}+\Delta^1_y(uv)^n_{i+1/2,j}$$

$$(F_y)^n_{i,j+1/2} = \Delta^1_x(uv)^n_{i,j+1/2}+\frac{v^n_{i,j+3/2}v^n_{i,j+1/2}-v^n_{i,j+1/2}v^n_{i,j-1/2}}{\Delta y}$$

$$VISX = \nu(\Delta_{yy}u^n_{i+1/2,j}-\Delta^1_y\Delta^1_x v^n_{i+1/2,j})$$

$$VIXY = \nu\Delta^1_x(\Delta^1_y u-\Delta^1_x v)^n_{i,j+1/2}$$

The discretized momentum equations used in SMAC present some differences with respect to MAC. One is the preference for the "*curl* of *curl*" form of the viscous term. The other is the so-called ZIP differencing of the cell-centered convective fluxes. This discretization ensures internal momentum conservation and removes part of a destabilizing truncation error that occurs in the original MAC discretization (Hirt, 1968, 1969). The first stage in the calculation consists of computing the temporary (tilde) velocities from (4.15) and (4.16). The second stage converts the $\tilde{u}$, $\tilde{v}$ field into the final velocity field so that $D^{n+1}_{i,j}=0$ for every cell. To preserve vorticity, the change must be given by the gradient of a potential function ψ, whose associated Poisson equation is (Amsden & Harlow, 1970)

$$D^{n+1}_{i,j} \equiv 0 = \tilde{D}_{i,j}-\Delta_{xx}\psi_{i,j}-\Delta_{yy}\psi_{i,j} \tag{4.17}$$

There is a clear similarity between (4.17) and the Poisson equation in the projection method. If $\theta_{i,j}=0$, the SMAC method becomes the projection method; and if $\theta_{i,j}$ satisfies the correct normal stress conditions on both free and rigid boundaries, then $\theta_{i,j}+\psi_{i,j}/\Delta t=\varphi_{i,j}$. The boundary conditions for $\psi_{i,j}$ are all homogeneous; for a rigid wall $\partial\psi/\partial n=0$, as in the projection method. This prescription leaves the velocity component normal to the wall unchanged (the treatment for an outflow boundary is sketched in figure 4.4) :

$$\begin{aligned}
\tilde{u}_{i-1/2,j} &= \tilde{u}_{i+1/2,j}\\
u_{i-1/2,j} &= \tilde{u}_{i-1/2,j}-\frac{1}{\Delta x}(\psi_{i,j}-\psi_{i-1,j})\\
\psi_{i-1,j} &= 0\\
v_{i-1,j+1/2} &= v_{i,j+1/2}
\end{aligned}$$

This is a combination of homogeneous Neumann boundary conditions for velocity with a homogeneous Dirichlet condition for ψ. In the first stage, homogeneous Neumann conditions are used; in the second stage (pseudo-pressure iteration) the velocities in the cells next to the wall are allowed to adjust to the changing ψ field, so that continuity is satisfied for these cells.

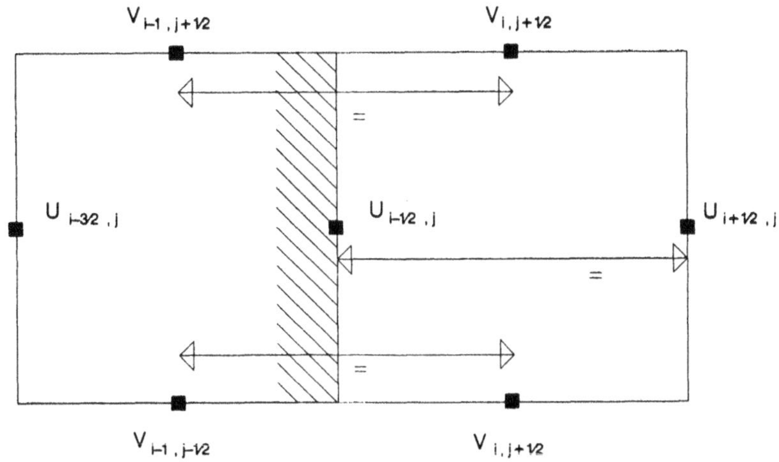

Figure 4.4 : *Outflow boundary conditions in SMAC*

The Poisson equation for the mass conservation (pseudo-pressure ψ) is solved by the same relaxation technique used in MAC, but the treatment of free-surfaces and internal obstacles is much simpler.

4.6 The pressure-velocity iteration method.

The pressure-velocity iteration method (Viecelli, 1969, 1971; Hirt et al. , 1975; Nichols et al. , 1980) is a highly simplified variant of the MAC method. The main difference with respect to other algorithms lies in the pressure solution, which does not require pressure boundary conditions to be formulated. The structure of the algorithm is as follows :

1) Explicit temporary (tilde) velocities are computed using the momentum equations in the x and y directions and the pressure gradient at time level n. This temporary velocity field has the correct vorticity at time level $n+1$, but in general does not satisfy the incompressibility condition.

2) Pressure and velocities are *simultaneously* iterated in each computational cell, so that the final velocity field is discretely divergence-free (up to some tolerance limit).

An example of an incompressible flow code based on the pressure-velocity iteration is the SOLA computer program (Hirt et al. , 1975). The temporary (tilde) velocities are computed explicitly by means of the following discretization

$$\tilde{u}_{i+1/2,j} = u^n_{i+1/2,j} + \Delta t[-\Delta^1_x \varphi^n_{i+1/2,j} - FUX - FUY + VISX + g_x] \tag{4.18}$$

$$\tilde{v}_{i,j+1/2} = v^n_{i,j+1/2} + \Delta t[-\Delta^1_y \varphi^n_{i,j+1/2} - FVX - FVY + VISY + g_y] \tag{4.19}$$

where

$$FUX = \Delta^1_x (u^2)^n_{i+1/2,j} + \frac{\alpha}{2}[|u^n_{i,j}|\Delta^-_x u^n_{i+1/2,j} - |u^n_{i+1,j}|\Delta^+_x u^n_{i+1/2,j}]$$

$$FUY = \Delta^1_y (uv)^n_{i+1/2,j} + \frac{\alpha}{2}[|v^n_{i+1/2,j-1/2}|\Delta^-_y u^n_{i+1/2,j} - |v^n_{i+1/2,j+1/2}|\Delta^+_y u^n_{i+1/2,j}]$$

$$FVX = \Delta^1_x (uv)^n_{i,j+1/2} + \frac{\alpha}{2}[|u^n_{i-1/2,j+1/2}|\Delta^-_x v^n_{i,j+1/2} - |u^n_{i+1,j+1/2}|\Delta^+_x v^n_{i,j+1/2}]$$

$$FVY = \Delta^1_y (v^2)^n_{i,j+1/2} + \frac{\alpha}{2}[|v^n_{i,j}|\Delta^-_y v^n_{i,j+1/2} - |v^n_{i,j+1}|\Delta^+_y v^n_{i,j+1/2}]$$

$$VISX = \nu \nabla^2_h u^n_{i+1/2,j}$$

$$VISY = \nu \nabla^2_h v^n_{i,j+1/2}$$

The convective terms are discretized using a combination of second-order central differences and first order upstream differencing. The coefficient α gives the desired amount of upstream differencing : $\alpha = 0$ for central differencing, $\alpha = 1$ for upstream differencing. Central differences are more accurate, but also unstable when combined with forward time differences and the viscosity is small (see § 4.8). For each particular problem, there will be an optimum α, which keeps the calculation stable without too much numerical diffusion.

The velocities computed according to (4.18)–(4.19) will not in general satisfy the continuity equation. The incompressibility constraint is then imposed by adjusting cell pressures according to their residual divergence, and using the momentum equation to compute velocity corrections induced by the pressure changes. This pressure adjustment must be done iteratively, because when the pressure in one cell is changed all its neighbors are affected. A heuristic derivation of the pressure-velocity iteration is given in § 5.3. More rigorous discussions of this topic can be found in Casulli & Greenspan (1988), and Viecelli (1969, 1971). A huge advantage of the pressure-velocity iteration technique is that there is no need for pressure boundary conditions, because the pressure correction in any cell is found without explicit reference to pressures in neighbor cells.

4.7 Numerical treatment of free-surfaces.

Free-surfaces pose three problems in numerical fluid dynamics : the surface must be numerically described; the surface must be advanced in time; and appropriate boundary conditions must be applied at the location of the surface. These three problems are related, because the algorithm used to advance the free-surface in time depends on the method used to define it, and the boundary conditions can only be applied after the location and shape of the surface are known.

4.7.1 Free-surface representation methods.

The main difficulty in this step is that the method should be able to treat any free-surface configuration (nearly-horizontal, nearly-vertical, bubbles, drops, overturning surfaces, etc) economically and without logical problems. Two types of methods have been used for defining the location and shape of free-surfaces: line/interface methods and region (or volume) methods. Examples of the first type of methods are the height function and line segment methods. Examples of region methods are the use of marker particles, and the so-called "volume of fluid" technique.

Height functions are the simplest method for treating free-surface problems. The free-surface is defined by a distance from a reference line. Thus for a free-surface that is nearly horizontal the reference line may be the bottom of the mesh and $\eta = \eta(x,t)$, the height above that datum, be approximated by a set of discrete values of η. In this approach, the slope of the surface $d\eta/dx$ must be smaller than the mesh aspect ratio $\Delta y/\Delta x$. Furthermore, the method does not work at all with multiple valued surfaces.

Line segments are a generalization of the height-function method. In this method, the free-surface is defined as a chain of short segments. These segments are defined by a set of ordered points whose coordinates are stored. The length of such segments should be smaller than the minimum cell size. More storage is required than in the previous case, but the method is not limited to single valued surfaces. There is one important difficulty: when surfaces intersect or when a surface folds over itself, segment chains must be reordered. Detection of such intersections and efficient reordering are difficult.

Thus, methods which define fluid regions rather than interfaces are advantageous in situations involving multiple free boundaries. This eliminates all ordering problems found in line/surface methods.

The first region method used was the Marker-particle (MAC) method (Welch et al. , 1966). Storage requirements increase significantly, because a large number of particle coordinates must be stored. It is also time-consuming to move all marker particles (an average of four to ten marker particles per cell is needed). A free-surface cell is defined as a cell containing marker particles, with at least one neighbor cell without marker particles. The actual location of the free-surface position within the cell is determined by an additional computation based on the distribution of marker particles within the cell.

Volume-of-fluid methods provide the same coarse interface information as the marker-particle method, but they require only one additional storage word for each mesh cell. Specifically, a function F is defined, whose value is unity at any point occupied by fluid and zero otherwise (Nichols & Hirt, 1975; Nichols et al. , 1980). The average value of F in a cell represents the fractional volume of the cell occupied by fluid : $F=1$ corresponds to a full cell and $F=0$ to an empty cell, whereas cells with $0<F<1$ and at least one empty neighbor cell are identified as free-surface cells. The storage requirements for F are identical to those for other dependent variables. Thus, the method is economical, and conceptually simple.

4.7.2 Methods for updating the free-surface.

The movement of a free-surface is described by the generic equation (kinematic free-surface boundary condition)

$$\frac{D\vec{x}_s}{Dt}=\vec{v}_s \tag{4.20}$$

where $\vec{x}_s$ is a point on the free-surface and $\vec{v}_s$ the velocity at the fluid surface. When the free-surface is described by the height function method, *i.e.* $\eta=\eta(x,t)$ (in a two-dimensional flow), the evolution equation for η is

$$\frac{\partial\eta}{\partial t}+u_s\frac{\partial\eta}{\partial x}=v_s \tag{4.21}$$

The velocity (u_s, v_s) at the free-surface is determined at the free-surface location by a simple local interpolation. When the free-surface is described by the cell-centered volume fraction function F, the basic equation for the convection of F is

$$\frac{\partial F}{\partial t}+u\frac{\partial F}{\partial x}+v\frac{\partial F}{\partial y}=0 \tag{4.22}$$

This equation simply states that F moves with the fluid, and is the PDE analogue of marker-particles. In other words, (4.22), is the equation which expresses the Lagrangian invariance of F, in an Eulerian representation. Because F is discontinuous, its derivatives and convective fluxes must be computed in a special way. In this work, the donor-acceptor flux approximation (Nichols & Hirt, 1975; Nichols et al. , 1980) is used to integrate (4.22) (see § 5.5.1 for a detailed discussion).

4.7.3 Discretization of free-surface boundary conditions.

After deciding which method will be used for free-surface representation and updating, it is necessary to impose the correct stress conditions at the free-surface location. For a two-dimensional surface, the normal and tangential stress conditions are

$$\varphi - 2\nu\left(n_x^2\frac{\partial u}{\partial x} + n_x n_y\left(\frac{\partial u}{\partial y} + \frac{\partial v}{\partial x}\right) + n_y^2\frac{\partial v}{\partial y}\right) = 0 \tag{4.23}$$

$$\nu\left(2n_x t_x\frac{\partial u}{\partial x} + (n_x t_y + n_y t_x)\left(\frac{\partial u}{\partial y} + \frac{\partial v}{\partial x}\right) + 2n_y t_y\frac{\partial v}{\partial y}\right) = 0 \tag{4.24}$$

where n_x and n_y are the components of the unit vector normal to the free-surface, and $t_x = n_y$ and $t_y = -n_x$ are the components of the tangential unit vector. In these equations it was assumed that the influence of the gas on the liquid is negligible. For a free-surface of small curvature these conditions can be replaced by the simpler expressions

$$\varphi - 2\nu\frac{\partial u_n}{\partial n} = 0 \tag{4.25}$$

$$\nu\left(\frac{\partial u_n}{\partial t} + \frac{\partial u_t}{\partial n}\right) = 0 \tag{4.26}$$

Usually, these simplified stress conditions are used in place of the complete stress conditions because the former are difficult to use and discretize on an arbitrary surface (Welch et al. , 1966; Hirt & Shannon, 1968; Nichols & Hirt , 1971).

Normal stress condition

The normal stress condition is imposed as a boundary condition for the pressure. Incorrect normal stress conditions result in a loss of momentum conservation in free-surface cells; the free-surface may then move too slowly or too fast (Hirt & Shannon, 1968), and the calculation may become unstable. The type of pressure boundary conditions in free-surface codes (MAC and SMAC), consisted of simply setting $\varphi_{i,j_{top}} = \varphi_s$, regardless of the position of the surface within the cell. This technique was not very satisfactory, however. An improved method consists of determining the free-surface position within the cell (Nichols & Hirt, 1971). Then, the pressure on the center of the ij_{top} cell is set in such a way that an interpolation between it and the adjacent full cell yields $\varphi = \varphi_s$ at the free-surface

$$\varphi_{i,j_{top}} = (1 - \eta_s)\varphi_{i,j_{top-1}} + \eta_s\varphi_s \tag{4.27}$$

with $\eta_s = d_c/d_{fs}$. This boundary condition is adequate if the fluid can be treated as inviscid. Otherwise an approximate free-surface pressure correction can be added to φ_s , and the discretized normal stress condition (for a nearly horizontal surface) becomes

$$\varphi_{i,j_{top}} = (1-\eta_s)\varphi_{i,j_{top-1}} + \eta_s(\varphi_s + \frac{2\nu}{\Delta y_j}(v_{i,j_{top+1/2}} - v_{i,j_{top-1/2}})) \tag{4.28}$$

A similar expression can be derived for a nearly vertical surface. This condition should be used only if the "cell Reynolds number" based on the typical velocity and cell size in the problem is sufficiently low (this question is clarified below).

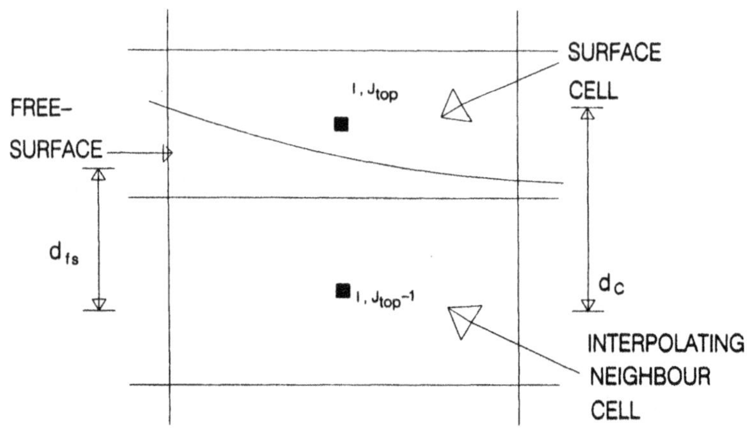

Figure 4.5 : *Definition sketch for the discretized normal stress condition*

Correct discretized versions of the tangential stress conditions are necessary in order to avoid a fictitious drag exerted on the fluid at surface cells, due to adjacent empty cells. Consequently, tangential stress conditions are imposed by setting the velocities at all boundaries between surface and empty cells using the continuity equation ($v_{i,j_{top+1/2}}$ in figure 4.6), and by setting velocities in empty cells adjacent to partial fluid cells ($u_{i+1/2,j_{top+1}}$ in figure 4.6) .

Tangential stress condition

According to figure 4.6 , we have

$$v_{i,j_{top+1/2}} = v_{i,j_{top-1/2}} - \frac{\Delta y_j}{\Delta x_i}(\Delta_x^1 u_{i,j_{top}})$$

using the discretized continuity equation for the surface cell. For the tangential stress conditions, two different solutions have been proposed

$$u_{i+1/2,jtop+1} = u_{i+1/2,jtop}$$

$$u_{i+1/2,jtop+1} = u_{i+1/2,jtop} - \frac{\Delta y_j}{\Delta x_i}(\Delta_x^1 v_{i+1/2,jtop+1/2})$$

The first is appropriate for inviscid flow calculations (Nichols & Hirt, 1971), while the second is appropriate when the viscosity is high (Amsden & Harlow, 1970). The decision on what type of conditions to use is made according to the cell Reynolds number, *i.e.* $max|\vec{u}|\Delta s/\nu$. This criterion is similar to the one used for deciding if the pressure correction should be included in the normal stress condition. Specifically, the inviscid boundary conditions should be used when the free-surface boundary layer is too thin to be resolved by the finite-difference mesh.

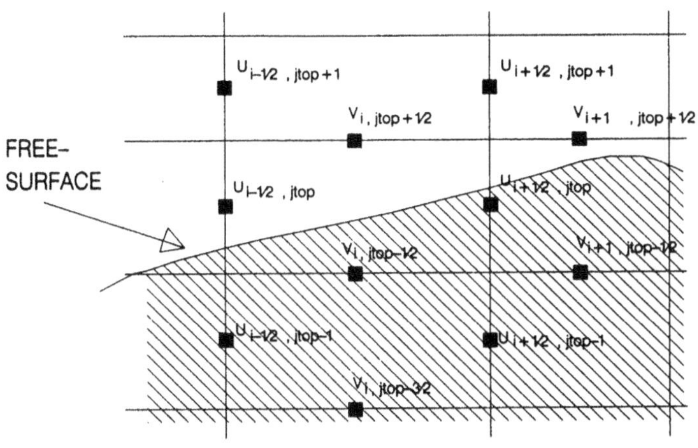

Figure 4.6 : *Definition sketch for the discretized tangential stress conditions.*

4.8 Stability considerations.

The subject of stability is very important, for it is strongly related with the general problem of convergence of the numerical solution. This relationship is expressed by Lax's equivalence theorem.

Consider the initial value problem for a generic vector property $\vec{f}(\vec{x},t)$

$$\frac{\partial \vec{f}}{\partial t} = \vec{P}(\vec{f}) \qquad \vec{f}(\vec{x},0) = \vec{f}_0 \quad given \tag{4.29}$$

and its discrete approximation

$$\vec{f}^{n+1}-\vec{f}^{n}=\Delta t[\vec{P}_1(\vec{f}^{n+1})+\vec{P}(\vec{f}^{n})]\ , \qquad t=\Delta t \qquad n=1,2,\ldots, \tag{4.30}$$

$\vec{P}$ is an operator which in general is nonlinear, and may also include side conditions (like the incompressibility constraint). $\vec{f}^n$ is defined at mesh points described by a vector of mesh intervals $\Delta\vec{x}$, and $\vec{P}_1$ and $\vec{P}_0$ are operators composed of polynomial combinations of difference operators such that $\vec{P}_0+\vec{P}_1$ is consistent with $\vec{P}$ (Morton, 1971). If (4.29) is a well-posed problem, (4.30) is said to be convergent if

$$\left\| m(\vec{f}^n)-(\vec{f})^n \right\| \to 0 \qquad or \qquad \left\| \vec{f}^n - M(\vec{f})^n \right\| \to 0$$

when Δt, $\Delta\vec{x}\to 0$, $n\Delta t\to t$ for all initial conditions (m and M are appropriate mappings between the function spaces on which $\vec{f}^n$ and $\vec{f}$ lie—Morton, 1971), possibly with some necessary restriction on the way in which $\Delta\vec{x}$ and Δt tend to zero.

The basic idea defining stability is that the numerical scheme, when applied exactly, should limit all components of the initial condition. In the Lax-Richtmeyer sense (Richtmeyer & Morton, 1967; Smith, 1985), a scheme is stable if there exists a positive constant K_1 such that

$$\left\| \vec{f}^n - \vec{w}^n \right\| \le K_1 \left\| \vec{f}^0 - \vec{w}^0 \right\|$$

uniformly while Δt, $\Delta\vec{x}\to 0$, $n\Delta t\to t$ for initial conditions $\vec{f}^0$ and $\vec{w}^0$. Stability may be conditional on some relationship between Δt and $\Delta\vec{x}$. The real importance of stability follows from the Lax equivalence theorem :

> "Given a well-posed linear initial value problem and a linear finite-difference approximation to it that satisfies the consistency condition, stability is the necessary and sufficient condition for convergence."

i.e. consistency+stability $\Rightarrow$ convergence. This theorem is discussed in depth in Richtmeyer & Morton (1967). Generalizations of Lax's theorem for nonlinear equations have been obtained by several authors (see Morton (1971) for a detailed discussion).

In practical situations, the Lax-Richtmeyer definition of stability is not very useful, because what is important is to assure boundedness of the solution for finite Δt and $\Delta\vec{x}$, and $n\Delta t\to\infty$. The difference between these two concepts of stability can be illustrated by considering the case $\vec{P}=$ *linear, constant coefficient operator*, and writing (4.30) in matrix form

$$\vec{f}^{n+1}=A\vec{f}^n+\vec{d}^n$$

where A is a square matrix and $\vec{d}^n$ is a vector which depends on the boundary conditions. The Lax-Richtmeyer definition implies (Smith, 1985)

$$\|A\| \leq 1 \tag{4.31}$$

whereas the "practical definition" conduces to (Smith, 1985)

$$\rho(A) = max|\lambda_i| \leq 1 \tag{4.32}$$

in which λ_i are the eigenvalues of A (in the Lax-Richtmeyer definition, A is not constant during the process of taking limits). In general (4.32) does not ensure convergence since $\rho(A) \leq \|A\| \leq 1$. Nevertheless, for practical purposes the gap between the two definitions is small, and (4.32) is the most useful stability result. A detailed discussion of the conditions in which (4.31) and (4.32) are equivalent can be found in Richtmeyer & Morton (1967) and Smith (1985).

Stability can be analyzed by means of the Fourier-von Neumann, energy, matrix, and Hirt's (truncation error) methods. In the Fourier-von Neumann method (Richtmeyer & Morton, 1967; Hindmarsh et al. 1984; Smith, 1985), the amplification properties of the discrete scheme are analyzed for a sinusoidal component which can be represented in the discrete mesh. This method is rigorous and sound, and extremely valuable because it tells not only how the components of the initial condition are amplified or attenuated, but also the speed at which they are propagated by the difference scheme. However, it is strictly valid only for linear problems (to which the principle of superposition applies) with periodic boundary conditions or in an infinite domain. The energy method is less systematic and precise (Morton, 1971), but can give rigorous results for a wider range of problems. The matrix method (Smith, 1985) is particularly adapted to the commonly used definition of stability (*i.e.* (4.32)), and has the advantage of taking into account the effect of the boundary conditions on stability. Hirt's stability theory (Hirt, 1968, 1969) is based on the idea that some part of the truncation error generates artificial "negative viscosity". This method is less formal and rigorous than von Neumann's method, but can be applied to both linear and nonlinear equations.

The objective of stability analyses is to find the maximum permissible Δt once a mesh $\Delta\vec{x}$ has been chosen. In the FD schemes used in MAC-type methods, instabilities may be generated in the momentum equations through the action of the convective and viscous terms, and in the pressure solution. It is often difficult to determine the origin of instability in actual calculations, for the relationship between all variables is complex.

Pressure-related instabilities are caused by incompatible pressure boundary conditions, or incorrect free-surface normal stress conditions, or the accumulation/growth of residual divergences in the mesh cells as the calculation progresses. The use of the pressure-velocity iteration technique with free-surface interpolation in the MAC-mesh greatly reduces the chance of pressure-triggered instabilities, because there is no need for pressure boundary conditions, and also because the pressure-velocity relaxation uses the residual divergence as the convergence criterion.

The greatest problems are posed by the explicit momentum approximation, whose stability can be analyzed as follows. In matrix notation, the FD approximations to the incompressible NS equations read (Bulgarelli et al. , 1984)

$$\vec{f}^{n+1} = \vec{F}(\vec{f}^n) - \Delta t A^{\mathrm{T}} \vec{\phi}^{n+1}$$

in which $\vec{f}^n$ is a vector whose elements are the velocities $u^n_{i+1/2,j}$ and $v^n_{i,j+1/2}$, and $\vec{\phi}^{n+1}$ is a vector whose elements are the pressures $\varphi^{n+1}_{i,j}$. A^{T} is the gradient matrix, and its transpose is the divergence matrix. The matrix version of the discretized PPE is

$$\vec{\phi}^{n+1} = \frac{1}{\Delta t}(AA^{\mathrm{T}})^{-1} A\vec{F}(\vec{f}^n)$$

from which it follows that

$$\vec{f}^{n+1} = [I - A^{\mathrm{T}}(AA^{\mathrm{T}})^{-1}A]\vec{F}(\vec{f}^n)$$

where I is the identity matrix and (AA^{T}) is invertible. Forcing the error $\vec{e}^{n+1}$ in $\vec{f}^n$ to be smaller than or equal to the error $\vec{e}^n$ in $\vec{f}^n$ in some suitable norm

$$\|\vec{e}^{n+1}\| \leq \|[I - A^{\mathrm{T}}(AA^{\mathrm{T}})^{-1}A][\vec{F}(\vec{f}^n + \vec{e}^n) - \vec{F}(\vec{f}^n)]\|$$

The matrix $[I - A^{\mathrm{T}}(AA^{\mathrm{T}})^{-1}A]$ is a projector, and therefore in the L_2 norm

$$\|\vec{e}^{n+1}\|_2 \leq \|[\vec{F}(\vec{f}^n + \vec{e}^n) - \vec{F}(\vec{f}^n)]\|_2 \tag{4.33}$$

Thus, the stability of the momentum equations can be analyzed independently of the pressure, provided that the projection of $\vec{f}^{n+1}$ onto the space of discrete divergence-free vector fields is correctly ensured. Because $\vec{F}(\vec{f}^n)$ is nonlinear in $\vec{f}^n$, it is still difficult to derive stability conditions from (4.33). If $\vec{F}(\vec{f}^n)$ is linearized, $\vec{F}(\vec{f}^n) = B(\vec{f}^n)\vec{f}^n + \vec{d}^n$ (where $B(\vec{f}^n)$ is a square matrix and $\vec{d}^n$ a known vector—Bulgarelli et al. , 1984) , and if it is assumed that $B(\vec{f}^n)$ is independent of $\vec{e}^n$, then (4.33) might be simplified to

$$\|\vec{e}^{n+1}\|_2 \leq \|B(\vec{f}^n)\|_2 \|\vec{e}^n\|_2 \tag{4.34}$$

(4.34) is similar to the conditions used to derive stability conditions for linear initial-value problems. Therefore, approximate stability criteria for the fluid flow equations can be obtained by analyzing a linear equation of the same type of the momentum equations. This linear equation is the convection-diffusion equation for a property $\vec{f}$

$$\frac{\partial \vec{f}}{\partial t} + \vec{u} \cdot \nabla \vec{f} = \nabla \cdot (\vec{\vec{K}}_f \cdot \nabla \vec{f}) \tag{4.35}$$

where $\vec{u}$ is a constant velocity and $\vec{\vec{K}}_f$ a constant diffusivity tensor. Stability conditions for the multi-dimensional version of (4.35) have been derived for the explicit schemes used in MAC-type methods (Hindmarsh et al. , 1984). These conditions will be used to select the most efficient alternative for discretizing the momentum equations in the present MAC-type model.

4.8.1 The convection-diffusion equation.

The basic model for analyzing the stability of the discretized momentum equations is the convection-diffusion equation

$$\frac{\partial f}{\partial t}+u\frac{\partial f}{\partial x}+v\frac{\partial f}{\partial y}=K_f\left(\frac{\partial^2 f}{\partial x^2}+\frac{\partial^2 f}{\partial y^2}\right) \tag{4.36}$$

where f is the dependent variable, and u, v and K_f have been assumed constant, and K_f is positive. For simplicity, the mesh is assumed uniform. For the purposes of the present work, the main interest is in 5-point explicit finite-difference schemes approximating (4.36). Three schemes are discussed : the FTCS (forward time, centered space) scheme, the first-order upwind scheme, and the modified FTCS scheme. Consider the one-dimensional version of (4.36)

$$\frac{\partial f}{\partial t}+u\frac{\partial f}{\partial x}=K_f\frac{\partial^2 f}{\partial x^2} \tag{4.36.a}$$

The FTCS approximation to (4.36.a) is

$$\frac{f_j^{n+1}-f_j^n}{\Delta t}+u\frac{f_{j+1}^n-f_{j-1}^n}{2\Delta x}=K_f\frac{f_{j+1}^n-2f_j^n-f_{j-1}^n}{\Delta x^2} \tag{4.37}$$

The Fourier-von Neumann method consists of examining the effect of (4.37) on a Fourier mode of the form $f=r^n exp(ijk\Delta x)$, for appropriate wave numbers k and associated wavelengths $\lambda=2\pi/|k|$ (i is the imaginary unit, $\surd(-1)$). Substituting into (4.37), it is found that the amplification factor, $r^{n+1}/r^n=r$, is equal to (Leonard, 1980)

$$r=1-\alpha_r(1-cos\theta)-iC_r sin\theta \tag{4.38}$$

with $\theta=k\Delta x$, $C_r=|u|\Delta t/\Delta x$ (the Courant number) and $\alpha_r=2K_f\Delta t/\Delta x^2$ (the diffusion parameter). For a stable calculation, $|r|\leq 1$, for $-\pi\leq\theta\leq\pi$. It can be shown that this requirement is equivalent to the pair of inequalities (Leonard, 1980)

$$C_r^2\leq\alpha_r\leq 1 \tag{4.39}$$

which in terms of the physical parameters of the problem is equivalent to the following limits for the maximum admissible Δt

$$\Delta t \leq \frac{\Delta x^2}{2K_f} \tag{4.40.a}$$

$$\Delta t \leq \frac{2K_f}{u^2} \tag{4.40.b}$$

The former is limiting when diffusion is dominant; the latter is the most restrictive condition in convection-dominated flows. The relative importance of diffusion and convection is measured by the grid Peclet number (Hindmarsh et al. , 1984)

$$Pe \equiv \frac{|u|\Delta x}{2K_f} = \frac{C_r}{\alpha_r}$$

$Pe<1$ corresponds to diffusion-dominated flows, and $Pe \geq 1$ to convection-dominated flows. The condition $C_r^2 \leq 1$ implies the so-called CFL (Courant-Friedrichs-Lewy) condition $C_r \leq 1$, *i.e.*

$$\Delta t \leq \frac{\Delta x}{|u|}$$

This condition is necessary, though not sufficient, for stability.

For the case of the 1-D convection-diffusion equation, the two stability conditions $C_r^2 \leq \alpha_r \leq 1$ can also be derived using Hirt's method (Hirt, 1968). Expanding each term in (4.37) in a Taylor series about the point (n,j) , it is found that the equation which is effectively being solved is

$$\frac{1}{2}\Delta t\frac{\partial^2 f}{\partial t^2}+\frac{\partial f}{\partial t}+u\frac{\partial f}{\partial x}=K_f\frac{\partial^2 f}{\partial x^2}+O(\Delta x^2,\Delta t^2) \tag{4.41}$$

The modified equation is hyperbolic, rather than parabolic (like the original equation), and propagates information into a region bounded by characteristic lines with slopes $\pm(2K_f/\Delta t)^{1/2}$ (in contrast with the original equation, which propagates information at an infinite speed). Requiring the mesh celerity $\Delta x/\Delta t$ to be inferior to the slope of these characteristic lines, gives the restriction $\alpha_r \leq 1$. To obtain the second stability condition using Hirt's method, it is necessary to express the derivative $\partial^2 f/\partial t^2$ in terms of spatial derivatives of f. This gives partial derivatives up to the fourth order in x, but if only the second order spatial derivative (which should be dominant) is kept, it is found that the modified equation is (Hirt, 1968)

$$\frac{\partial f}{\partial t}+u\frac{\partial f}{\partial x}=(K_f-\frac{1}{2}\Delta t u^2)\frac{\partial^2 f}{\partial x^2}+O(\Delta x^2,\Delta t^2) \tag{4.42}$$

Thus, it is seen that the combination FT–CS introduces a negative diffusion coefficient. It is the FT, not the CS part of the scheme, that is responsible for this negative diffusion coefficient. For stable solutions, $\Delta t \leq 2K_f/u^2$, recovering the second condition obtained using the Fourier-von Neumann method.

The FTCS scheme can be very restrictive for simulating convection- dominated flows. For this reason, the convective term may be discretized using the first order upwind scheme

$$\frac{f_j^{n+1}-f_j^n}{\Delta t}+\frac{1}{2}u[(1-s)\frac{f_{j+1}^n-f_j^n}{\Delta x}+(1+s)\frac{f_j^n-f_{j-1}^n}{\Delta x}]=K_f\frac{f_{j+1}^n-2f_j^n-f_{j-1}^n}{\Delta x^2} \tag{4.43}$$

with $s=sign(u)$. This scheme compensates the inherent negative diffusion coefficient induced by the FT part of (4.43) with a positive diffusion coefficient induced by the discretization of the convective term. This procedure is not very reliable, because it masks a truncation error originated in one term with another truncation error of different origin.

The effective diffusion coefficient for first order upwind is $K_f+|u|\Delta x/2$. Repeating the Fourier-von Neumann analysis for the FTCS method with K_f replaced by this effective diffusion coefficient, it is found that (Hindmarsh et al. , 1984)

$$\Delta t \leq \frac{\Delta x}{|u|} \tag{4.44}$$

$$\Delta t \leq \frac{1}{(\frac{2K_f}{\Delta x^2}+\frac{|u|}{\Delta x})} \tag{4.45}$$

The last condition is much less restrictive than the FTCS condition for convection-dominated flows. Furthermore, the scheme can still be used in the limit case $K=0$. However, the numerical diffusion may cause gross inaccuracies in the computed solution.

An alternative which combines the virtues of FTCS and first-order upwind is the modified FTCS scheme (Hindmarsh et al. , 1984; Roache, 1972)

$$\frac{f_j^{n+1}-f_j^n}{\Delta t}+u\frac{f_{j+1}^n-f_{j-1}^n}{2\Delta x}=(K_f+\frac{1}{2}u^2\Delta t)\frac{f_{j+1}^n-2f_j^n-f_{j-1}^n}{\Delta x^2} \tag{4.46}$$

This scheme "corrects" the most destructive truncation error (the negative viscosity error) by replacing K_f by $K_f+\frac{1}{2}u^2\Delta t$; it has much better properties than either FTCS or upwind. Introducing the effective viscosity in the Fourier-von Neumann stability analysis for the FTCS scheme, it is found that the stability constraints are $C_r^2 \leq \alpha_r + C_r^2 \leq 1$. These give $\alpha_r \geq 0$ and

$$\Delta t \leq \frac{K_f}{u^2}\{(1+(\frac{u\Delta x}{K_f})^2)^{1/2}-1\} \tag{4.47}$$

The 1-D modified FTCS scheme is stable for $K_f=0$, if $C_r \leq 1$.

Stability analyses of the two- and three-dimensional convection-diffusion equation are much more involved than those for the one-dimensional version. Recently, a number of stability results for the multi-dimensional cases has been published (*e.g.* Hindmarsh et al. , 1984). For the two-dimensional FTCS scheme, the constraints on the maximum allowable Δt are (Hindmarsh et al. , 1984)

$$\Delta t \le \frac{\Delta x^2 \Delta y^2}{2K(\Delta x^2 + \Delta y^2)} \tag{4.48}$$

$$\Delta t \le \frac{2K_f}{u^2 + v^2} \tag{4.49}$$

$$\Delta t \le min\left\{\frac{\Delta x}{|u|}, \frac{\Delta y}{|v|}\right\} \tag{4.50}$$

For the two-dimensional upwind scheme, the limiting Δt should satisfy the CFL condition, and the following inequality (Hindmarsh et al. , 1984)

$$\Delta t \le \frac{1}{\left(\frac{2K_f}{\Delta x^2} + \frac{2K_f}{\Delta y^2} + \frac{|u|}{\Delta x} + \frac{|v|}{\Delta y}\right)} \tag{4.51}$$

For the modified FTCS scheme in two dimensions, it is found that the limiting Δt should satisfy the following inequalities (Hindmarsh et al. , 1984)

$$\Delta t \le \frac{\{(K_f(\frac{1}{\Delta x^2} + \frac{1}{\Delta y^2}))^2 + (\frac{u}{\Delta x})^2 + (\frac{v}{\Delta y})^2\}^{1/2} - K_f(\frac{1}{\Delta x^2} + \frac{1}{\Delta y^2})}{(\frac{u}{\Delta x})^2 + (\frac{v}{\Delta y})^2} \tag{4.52}$$

$$\Delta t \le \frac{2K_f}{|uv|} \tag{4.53}$$

The last of these does not apply when either u or v is zero. In contrast to the 1-D case, the modified FTCS scheme is not stable in two dimensions when $K_f = 0$. Nevertheless, this scheme still has advantages with respect to FTCS.

4.8.2 Stability of the momentum equations.

Making $f = u$, v and $K_f = \nu$, the stability limits for the 2-D convection-diffusion equation can be transposed to the momentum equations. However, these limits are based on a linear analysis which neglects the pressure gradient and the influence of boundary conditions.

Stability of the nonlinear equations can be studied using Hirt's method (or truncation-error method). In this method, all terms in the discretized equations are expanded in Taylor series, keeping only the lowest-order even and odd derivative terms. The limiting Δt is then found by requiring that the "most destructing" truncation error, usually associated with a negative artificial viscosity, should not outweigh the physical diffusion term. Using this method, Hirt (1968, 1969) showed that for the original MAC discretization the equations which are *effectively* solved are

$$\frac{\partial u}{\partial t}+u\frac{\partial u}{\partial x}+v\frac{\partial u}{\partial y}=-\frac{\partial \varphi}{\partial x}+(\nu-\frac{1}{2}\Delta t u^2-\frac{\Delta x^2}{2}\frac{\partial u}{\partial x})\frac{\partial^2 u}{\partial x^2}$$
$$+(\nu-\frac{1}{2}\Delta t v^2-\frac{\Delta y^2}{4}\frac{\partial v}{\partial y})\frac{\partial^2 u}{\partial y^2} \tag{4.54}$$

$$\frac{\partial v}{\partial t}+u\frac{\partial v}{\partial x}+v\frac{\partial v}{\partial y}=-\frac{\partial \varphi}{\partial y}+(\nu-\frac{1}{2}\Delta t u^2-\frac{\Delta x^2}{4}\frac{\partial u}{\partial x})\frac{\partial^2 v}{\partial x^2}$$
$$+(\nu-\frac{1}{2}\Delta t v^2-\frac{\Delta y^2}{2}\frac{\partial v}{\partial y})\frac{\partial^2 v}{\partial y^2} \tag{4.55}$$

$$\frac{\partial u}{\partial x}+\frac{\partial v}{\partial y}=0 \tag{4.56}$$

Thus, the destabilizing truncation error (acting as a negative viscosity term) has two contributions: one proportional to Δt, the other to Δx^2 , Δy^2 and gradients of velocity. The first can be removed by reducing the time step, but the second can only be avoided by increasing mesh resolution.

According to Hirt's analysis, the stability conditions for the MAC discretization are (Hirt, 1968; Hirt & Cook, 1972)

$$\Delta t \leq max\{\frac{2\nu}{u^2}, \frac{2\nu}{v^2}\} \tag{4.57}$$

$$\Delta x^2 \leq 2\nu / \left|\frac{\partial u}{\partial x}\right| \tag{4.58}$$

$$\Delta y^2 \leq 2\nu / \left|\frac{\partial v}{\partial y}\right| \tag{4.59}$$

4.9 Conclusions.

In order to solve the central problem addressed in this work, as defined in chapter 3, we need an algorithm with the following characteristics: ability to treat any free-surface configuration efficiently; simplicity of both momentum and pressure solutions; and flexibility with respect to changes during code development (such as testing of different discretizations, introduction of new terms, incorporation of additional equations, etc). From the comparative review here presented, it is possible to define the structure of a basic algorithm that meets these requirements. The technical aspects of such an algorithm are summarized in table 4.1 .

Table 4.1 : Basic characteristics of the selected solution algorithm.

Computational aspect	Selected approach
Momentum equation approximation	Combined modified FTCS/Upwind scheme
Continuity equation approximation	Pressure-velocity iteration
Free-surface definition and updating	VOF method
Free surface boundary conditions :	
Normal stress	Viscous/inviscid, with pressure interpolation
Tangential stress	Viscous/inviscid
Mesh	Variable-size MAC mesh, with provision for simple internal obstacles

Chapter 5

DESCRIPTION
OF THE
NUMERICAL MODEL

5.1 Introduction.

This chapter describes a computer program, 2D-HYDROTUR, developed for the specific purpose of studying free-surface breaking flows. The features of 2D-HYDROTUR include: two-dimensional laminar or turbulent flow; rectangular mesh of variable size; free-surface or confined flows; free-slip, no-slip, law-of-the-wall, and free-outflow boundary conditions at regular mesh boundaries; viscous and inviscid free-surface boundary conditions; internal obstacles of arbitrary shape, bounded by mesh cells; automatic time step control and cycling; and built-in and off-line stability estimates.

The governing equations are the two-dimensional momentum and continuity equations, the volume-of-fluid evolution equation, and the modeled K and ϵ equations (if the turbulence model is "on")

$$\frac{\partial U}{\partial t}+U\frac{\partial U}{\partial x}+V\frac{\partial U}{\partial y}=-\frac{\partial \varphi}{\partial x}+\nu(\frac{\partial^2 U}{\partial x^2}+\frac{\partial^2 U}{\partial y^2})+g_x+$$

$$+\boldsymbol{\frac{\partial}{\partial x}(2\nu_t\frac{\partial U}{\partial x})+\frac{\partial}{\partial y}(\nu_t(\frac{\partial U}{\partial y}+\frac{\partial V}{\partial x}))} \tag{5.1}$$

$$\frac{\partial V}{\partial t}+U\frac{\partial V}{\partial x}+V\frac{\partial V}{\partial y}=-\frac{\partial \varphi}{\partial y}+\nu(\frac{\partial^2 V}{\partial x^2}+\frac{\partial^2 V}{\partial y^2})+g_y+$$

$$+\boldsymbol{\frac{\partial}{\partial x}(\nu_t(\frac{\partial U}{\partial y}+\frac{\partial V}{\partial x}))+\frac{\partial}{\partial y}(2\nu_t(\frac{\partial V}{\partial y}))} \tag{5.2}$$

$$\frac{\partial U}{\partial x}+\frac{\partial V}{\partial y}=0 \tag{5.3}$$

$$\frac{\partial F}{\partial t}+U\frac{\partial F}{\partial x}+V\frac{\partial F}{\partial y}=0 \tag{5.4}$$

$$\boldsymbol{\frac{\partial K}{\partial t}+U\frac{\partial K}{\partial x}+V\frac{\partial K}{\partial y}=\frac{\partial}{\partial x}((\nu+\frac{\nu_t}{\sigma_k})\frac{\partial K}{\partial x})+\frac{\partial}{\partial y}((\nu+\frac{\nu_t}{\sigma_k})\frac{\partial K}{\partial y})+\mathcal{P}-\epsilon} \tag{5.5}$$

$$\boldsymbol{\frac{\partial \epsilon}{\partial t}+U\frac{\partial \epsilon}{\partial x}+V\frac{\partial \epsilon}{\partial y}=\frac{\partial}{\partial x}((\nu+\frac{\nu_t}{\sigma_\epsilon})\frac{\partial \epsilon}{\partial x})+\frac{\partial}{\partial y}((\nu+\frac{\nu_t}{\sigma_\epsilon})\frac{\partial \epsilon}{\partial y})+C_{1\epsilon}\frac{\epsilon}{K}\mathcal{P}-C_{2\epsilon}\frac{\epsilon^2}{K}} \tag{5.6}$$

with $\mathcal{P}=\nu_t\{2(\frac{\partial U}{\partial x})^2+2(\frac{\partial V}{\partial y})^2+(\frac{\partial U}{\partial y}+\frac{\partial V}{\partial x})^2\}$ and $\nu_t=C_d\frac{K^2}{\epsilon}$.

If a laminar flow is to be simulated, the dependent variables are the instantaneous horizontal and vertical velocity components, pressure (divided by constant density) and volume fraction, *i.e.* U, V, φ, and F; the bold-faced terms (modeled Reynolds stresses) and equations (K-ϵ) in (5.1)–(5.6) are not computed. If a turbulent flow is to be simulated, U, V, φ, and F are interpreted as the mean flow velocity components, mean pressure (divided by constant density), and volume fraction, respectively. In this case, the bold-faced terms and equations are included in the solution.

Equations (5.1)–(5.6) are appropriate for both free-surface and confined flows. In the latter case, the condition $F(\vec{x},0)=1$,$\forall\vec{x}\in\Omega$ and the boundary conditions for F on Σ assure that $F(\vec{x},t)=1 \quad \forall\vec{x}\in\Omega$, $t>0$. These equations are discretized on an Eulerian mesh of rectangular cells with variable size (figure 5.1) . The computational domain is surrounded by a layer of fictitious cells. These cells are used to set velocity boundary conditions, so that the same discretized equations are used in the whole computational domain. The location of the dependent variables in each mesh cell is represented in figure 5.2 . The grid layout employed is not as flexible as a mesh composed of arbitrary quadrilaterals. However, it gives a considerable advantage for localized resolution over methods based on uniform meshes.

The basic computational cycle is the following :

1) Explicit approximations of the momentum equations are used to compute a first guess for the new time-level ($n+1$) velocities, using the previous (time level n) values for the convective, viscous and pressure gradient accelerations. The modeled Reynolds stresses are added in a separate subprogram, which is bypassed if the turbulence is "off".

2) To satisfy the continuity equation, pressures are iteratively adjusted in each cell; velocity changes induced by the pressure corrections are computed, and the residual divergence in each cell is driven to zero.

3) If the turbulence model is "on", the modeled K and ϵ equations are used to advance K and ϵ to the new time level. The updated values of K and ϵ are used to compute ν_t , which becomes available for the next explicit momentum approximation.

4) The F function, defining fluid-occupied regions, is updated to give the new (time level $n+1$) fluid configuration. All mesh cells are reflagged as full, surface or empty.

5) All variables are updated ($n \leftarrow n+1$), the time and cycle counters are incremented, and the cycle is restarted.

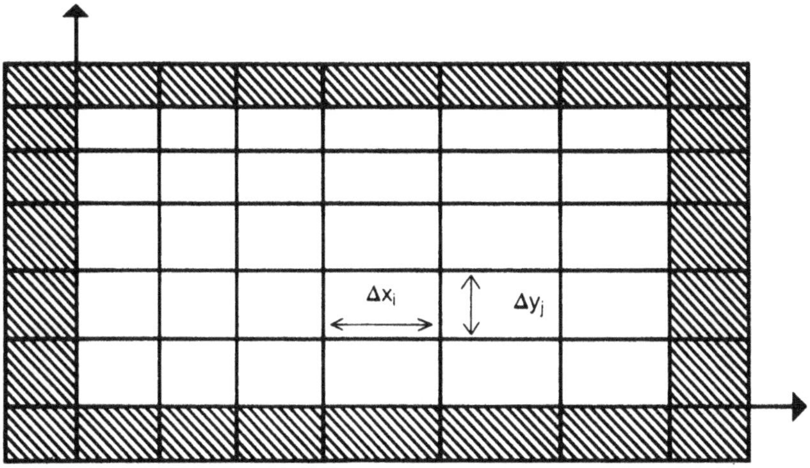

Figure 5.1 : *Finite-difference mesh with variable rectangular cells.*

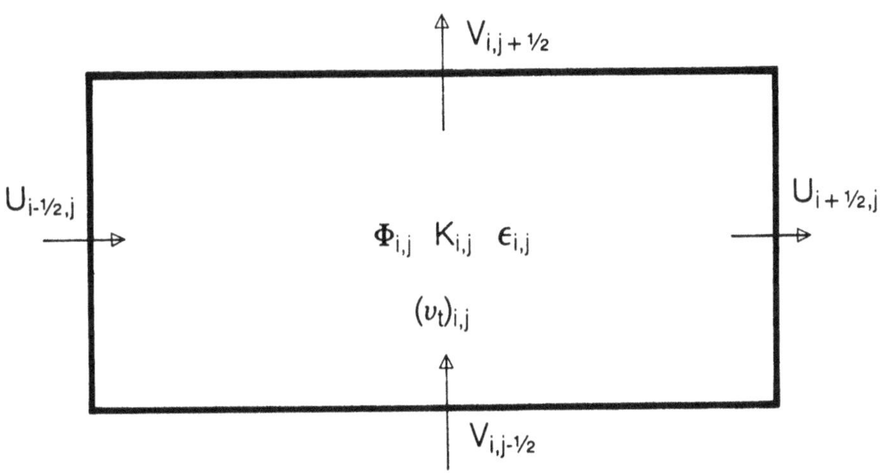

Figure 5.2 : *Location of variables in a typical mesh cell.*

The structure chart of the 2D-HYDROTUR program is presented in figure 5.3.

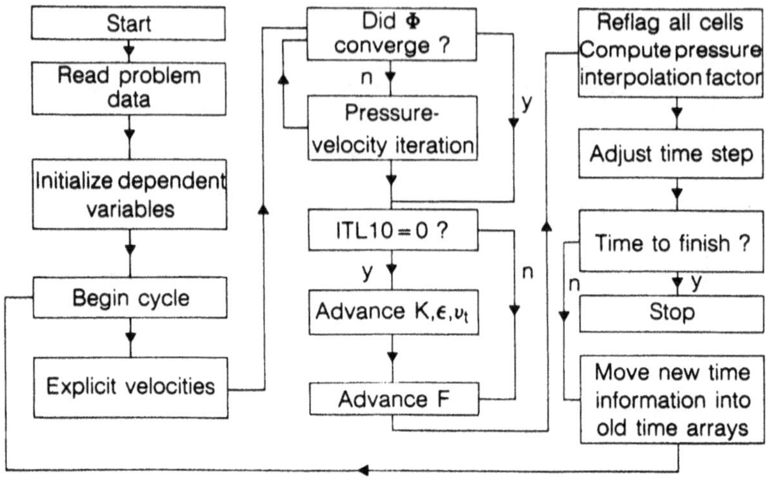

Figure 5.3 : *Structure chart of the 2D−HYDROTUR computer program.*

5.2 Momentum equation approximations.

A generic finite-difference approximation to the momentum equations is

$$\tilde{U}_{i+1/2,j}=U^n_{i+1/2,j}+\Delta t[\ -\frac{\varphi^n_{i+1,j}-\varphi^n_{i,j}}{\Delta x_{i+1/2}}-FUX^n_{i+1/2,j}-FUY^n_{i+1/2,j}$$

$$+VISX^n_{i+1/2,j}+g_x+\iota REYNX^n_{i+1/2,j}] \tag{5.7}$$

$$\tilde{V}_{i,j+1/2}=V^n_{i,j+1/2}+\Delta t[\ -\frac{\varphi^n_{i,j+1}-\varphi^n_{i,j}}{\Delta y_{j+1/2}}-FVX^n_{i,j+1/2}-FVY^n_{i,j+1/2}$$

$$+VISY^n_{i,j+1/2}+g_y+\iota REYNY^n_{i,j+1/2}] \tag{5.8}$$

In (5.7)–(5.8) , FUX, FUY, FVX, and FVY are approximations of the convective accelerations $U(\partial U/\partial x)$, $V(\partial U/\partial y)$, $U(\partial V/\partial x)$ and $V(\partial V/\partial y)$, respectively; $VISX$ and $VISY$ are approximations of the viscous accelerations in the x and y directions; $REYNX$ and $REYNY$ are approximations to the modeled Reynolds-stress accelerations in the x and y directions; and ι is an indicator whose value is 1 for turbulent flows and 0 for laminar flows. Following Nichols et al. (1980), the convective terms are discretized using a combination of first-order upwind and central difference approximations, with a parameter α that controls the relative amount of each one. The expression for $FUX^n_{i+1/2,j}$ is

$$FUX^n_{i+1/2,j}=\frac{U^n_{i+1/2,j}}{\Delta x_\alpha}[\Delta x_{i+1}DUL+\Delta x_i DUR+$$

$$+\alpha sign(U^n_{i+1/2,j})(\Delta x_{i+1}DUL-\Delta x_i DUR)] \tag{5.9}$$

$$DUL=\Delta_x^- U^n_{i+1/2,j}$$
$$DUR=\Delta_x^+ U^n_{i+1/2,j}$$
$$\Delta x_\alpha=\Delta x_{i+1}+\Delta x_i+\alpha sign(U^n_{i+1/2,j})(\Delta x_{i+1}-\Delta x_i)$$

When $\alpha=0$, this expression gives the second-order accurate centered-difference approximation; when $\alpha=1$, the first-order upwind approximation is recovered. In the same way, the expression for $FUY^n_{i+1/2,j}$ is

$$FUY^n_{i+1/2,j}=\frac{V^n_{i+1/2,j}}{\Delta y_\alpha}[\Delta y_b DUT+\Delta y_t DUB+$$

$$+\alpha sign(V^n_{i+1/2,j})(\Delta y_t DUB-\Delta y_b DUT)] \tag{5.10}$$

$$DUT=\Delta_y^+ U^n_{i+1/2,j}$$
$$DUB=\Delta_x^- U^n_{i+1/2,j}$$

$$\Delta y_b=\frac{\Delta y_j+\Delta y_{j-1}}{2}=\Delta y_{j-1/2}$$

$$\Delta y_t=\frac{\Delta y_{j+1}+\Delta y_j}{2}=\Delta y_{j+1/2}$$

$$\Delta y_\alpha=\Delta y_t+\Delta y_b+\alpha sign(V^n_{i+1/2,j})(\Delta y_t-\Delta y_b)$$

The approximations for the convective accelerations in the y direction are

$$FVX^n_{i,j+1/2}=\frac{U^n_{i,j+1/2}}{\Delta x_\alpha}[\Delta x_l DVR+\Delta x_r DVL+$$

$$+\alpha sign(U^n_{i,j+1/2})(\Delta x_r DVL-\Delta x_l DVR)] \tag{5.11}$$

$$FVY^{n+1}_{i,j+1/2}=\frac{V^n_{i,j+1/2}}{\Delta y_\alpha}[\Delta y_j DVT+\Delta y_{j+1}DVB+$$

$$+\alpha sign(V^n_{i,j+1/2})(\Delta y_{j+1}DVB-\Delta y_j DVT)] \tag{5.12}$$

$$DVR=\Delta_x^+ V^n_{i,j+1/2}$$
$$DVL=\Delta_x^- V^n_{i,j+1/2}$$

$$DVT = \Delta_y^+ V_{i,j+1/2}^n$$
$$DVB = \Delta_y^- V_{i,j+1/2}^n$$

$$\Delta x_l = \frac{\Delta x_i + \Delta x_{i-1}}{2} = \Delta x_{i-1/2}$$

$$\Delta x_r = \frac{\Delta x_{i+1} + \Delta x_i}{2} = \Delta x_{i+1/2}$$

$$\Delta x_\alpha = \Delta x_l + \Delta x_r + \alpha sign(U_{i,j+1/2}^n)(\Delta x_r - \Delta x_l)$$
$$\Delta y_\alpha = \Delta y_j + \Delta y_{j+1} + \alpha sign(V_{i,j+1/2}^n)(\Delta y_{j+1} - \Delta y_j)$$

In the present work, the following two rules are observed :

1) When the quantities in the finite-difference expressions are required at positions where they are not defined (*e.g.* $V_{i+1/2,j}^n$ in $FUY_{i+1/2,j}^n$, $U_{i,j+1/2}^n$ in $FVX_{i,j+1/2}^n$, etc), they are interpolated before using them in the finite-difference equations.

2) When a product between any quantities is required, the quantities are averaged (interpolated) before the product is formed.

The basic idea used in the discretization of the convective terms is to weight the upstream and downstream derivatives by the factors $1+\alpha$ and $1-\alpha$ respectively. The derivatives are also weighted by cell sizes in order to maintain formal accuracy in the variable mesh.

The viscous acceleration terms are approximated by standard second-order central differences, plus a local "balancing tensor viscosity" term that corrects the most destabilizing truncation error arising from the use of forward time differences. Neglecting the effect of the pressure gradient, the modified momentum equation arising from the FT discretization is

$$\frac{\Delta t}{2}\frac{\partial^2 \vec{U}}{\partial t^2} + \frac{\partial \vec{U}}{\partial t} = -\nabla\varphi - \vec{U}\cdot\nabla\vec{U} + \nu\nabla^2\vec{U} + \vec{g}$$

To derive the expression for the local truncation error originated by the FT discretization, the first term on the left hand side of this equation is expressed in terms of spatial derivatives of $\vec{U}$. Writing the convection-diffusion part in operator form and keeping the lowest order spatial derivatives, it is found that

$$\frac{\partial^2}{\partial t^2} = U^2\frac{\partial^2}{\partial x^2} + V^2\frac{\partial^2}{\partial y^2} + 2UV\frac{\partial^2}{\partial x\partial y} + U\left(\frac{\partial U}{\partial x}\frac{\partial}{\partial x} + \frac{\partial V}{\partial x}\frac{\partial}{\partial y}\right) + V\left(\frac{\partial U}{\partial y}\frac{\partial}{\partial x} + \frac{\partial V}{\partial y}\frac{\partial}{\partial y}\right)$$

Substituting this expression into the modified momentum equations, and retaining only the diffusion-like truncation error, it is found that

$$\frac{\partial U}{\partial t}+U\frac{\partial U}{\partial x}+V\frac{\partial U}{\partial y} = -\frac{\partial \varphi}{\partial x}+(\nu-\frac{\Delta t}{2}U^2)\frac{\partial^2 U}{\partial x^2}+(\nu-\frac{\Delta t}{2}V^2)\frac{\partial^2 U}{\partial y^2}+g_x$$

$$\frac{\partial V}{\partial t}+U\frac{\partial V}{\partial x}+V\frac{\partial V}{\partial y} = -\frac{\partial \varphi}{\partial y}+(\nu-\frac{\Delta t}{2}U^2)\frac{\partial^2 V}{\partial x^2}+(\nu-\frac{\Delta t}{2}V^2)\frac{\partial^2 V}{\partial y^2}+g_y$$

i.e. the FT part of the discretization induces a local negative viscosity. This destabilizing truncation error can be corrected by adding a "balancing tensor viscosity" to the viscous acceleration term (in the spirit of the modified FTCS scheme). Using this idea, the following discretization of the viscous acceleration term was adopted

$$VISX = [\nu+(1-\alpha)\frac{\Delta t}{2}(U^2)^n_{i+1/2,j}]\Delta_{xx}U^n_{i+1/2,j}+[\nu+(1-\alpha)\frac{\Delta t}{2}(V^2)^n_{i+1/2,j}]\Delta_{yy}U^n_{i+1/2,j}$$

$$VISY = [\nu+(1-\alpha)\frac{\Delta t}{2}(U^2)^n_{i,j+1/2}]\Delta_{xx}V^n_{i,j+1/2}+[\nu+(1-\alpha)\frac{\Delta t}{2}(V^2)^n_{i,j+1/2}]\Delta_{yy}V^n_{i,j+1/2}$$

The factor $1-\alpha$ ensures that the amount of "balancing tensor viscosity" equals the amount of central differencing used in the calculation. Otherwise, when upwind differencing is used for the convective term, the two smoothing effects would be superposed in an undesirable way. It was found that the modified FTCS/balancing tensor viscosity scheme is less destructive than any combination including first order upwind; this latter is provided in the code as an additional degree of freedom. It is also possible to use central differencing without any smoothing, if the user so desires.

The full expressions for the modeled Reynolds stresses accelerations were used in some test problems, but were not implemented in the final version of the program. The following simplified closures were used instead

$$\nu_t(\frac{\partial^2 U}{\partial x^2}+\frac{\partial^2 U}{\partial y^2}) \qquad (x\text{-}momentum)$$

$$\nu_t(\frac{\partial^2 V}{\partial x^2}+\frac{\partial^2 V}{\partial y^2}) \qquad (y\text{-}momentum)$$

The full expressions generated instabilities and errors in some free-surface simulations. These were eliminated by the use of the simplified expressions, possibly because in that case the number of variable-mesh interpolations is much smaller. The discrete approximations for the modeled Reynolds stresses are

$$REYNX = (\nu_t)^n_{i+1/2,j}(\Delta_{xx}+\Delta_{yy})U^n_{i+1/2,j}$$
$$REYNY = (\nu_t)^n_{i,j+1/2}(\Delta_{xx}+\Delta_{yy})V^n_{i,j+1/2}$$

Following the usual convention, $(\nu_t)^n_{i+1/2,j}$ and $(\nu_t)^n_{i,j+1/2}$ are interpolated before they are used in the finite-difference approximations.

Remarks

1. The "balancing tensor diffusivity" was introduced in numerical fluid dynamics by Dukowicz & Ramshaw (1979), using the conservative form for the convection term. Consequently, their expression for the balancing viscosity term is different from the one presented above.

2. According to recent evidence (*e.g.* Hindmarsh et al. , 1984), the usefulness of the truncation error correction/modified FTCS method is essentially independent of the conservative property, type of flow (compressible or incompressible—Dukowicz & Ramshaw, 1979), and even type of discretization (FD or FE—Gresho et al. , 1984). Due to its basic simplicity, modified FTCS seems to be competitive with other schemes, including flux-corrected transport (Boris & Book, 1976; Zalesak, 1979; Smolarkiewicz, 1984).

3. In the computer program, the explicit momentum approximations are not computed in the following conditions :

- If $F_{i,j}+F_{i+1,j}<\epsilon_f$, $\tilde{U}$ is not computed and set to zero; in the same way, if $F_{i,j}+F_{i,j+1}<\epsilon_f$, $\tilde{V}$ is not computed and set to zero. ϵ_f is a small value, typically 10^{-6}, used to negate round-off error effects when testing for $F=1$ or $F=0$ (§ 5.5.1).

- If one of the cells i,j or $i+1,j$ is an obstacle cell, $\tilde{U}$ is not computed and set to zero; If one of the cells i,j or $i,j+1$ is an obstacle cell, $\tilde{V}$ is not computed and set to zero. This ensures impermeability of the obstacles.

4. To improve the performance of the K-ϵ model for describing low-intensity turbulence, the eddy-viscosity coefficient is multiplied by the damping function (see § 2.9.4)

$$\frac{\nu_t}{\beta\nu}[1-exp(-\beta\frac{\nu}{\nu_t})]$$

where $\beta\simeq 100$. This is also done in the numerical approximations of the diffusion terms in the K and ϵ equations (§ 5.4).

5.3 Continuity equation approximation.

Velocities computed from equations (5.7)–(5.8) will not in general satisfy the continuity equation because φ^{n+1} is not available. To satisfy the continuity equation and determine the correct pressure, pressures and velocities must be adjusted in each cell occupied by fluid. In all full cells, pressure is changed in such a way that the divergence $D_{i,j}$ left by the first step is driven to zero; in all free-surface cells, the cell pressure is determined in such a way that a linear interpolation between the pressures in the surface and in the adjacent full cell yields the wanted value φ_s (usually zero) at the free-surface location. In both cases, the velocities located on the sides of the cell are simultaneously adjusted, in response to the pressure change in the cell.

Pressure iteration in a full cell

The new estimate for the cell pressure is

$$\varphi_{i,j}^{(n+1)} \leftarrow \varphi_{i,j}^{(n)} + \Delta\varphi_{i,j}$$

where $^{(n+1)}$ and $^{(n)}$ denote iteration levels. From (5.7)–(5.8), it is found that the new estimates for the velocities on the sides of the i,j cell are

$$U_{i+1/2,j}^{(n+1)} \leftarrow U_{i+1/2,j}^{(n)} + \frac{\Delta t}{\Delta x_{i+1/2}} \Delta\varphi_{i,j} \tag{5.13.a}$$

$$U_{i-1/2,j}^{(n+1)} \leftarrow U_{i-1/2,j}^{(n)} - \frac{\Delta t}{\Delta x_{i-1/2}} \Delta\varphi_{i,j} \tag{5.13.b}$$

$$V_{i,j+1/2}^{(n+1)} \leftarrow V_{i,j+1/2}^{(n)} + \frac{\Delta t}{\Delta y_{j+1/2}} \Delta\varphi_{i,j} \tag{5.13.c}$$

$$V_{i,j-1/2}^{(n+1)} \leftarrow V_{i,j-1/2}^{(n)} - \frac{\Delta t}{\Delta y_{j-1/2}} \Delta\varphi_{i,j} \tag{5.13.d}$$

where the superscripts $^{(n+1)}$ and $^{(n)}$ denote iteration levels. The pressure correction $\Delta\varphi_{i,j}$ is computed from the requirement $D_{i,j}^{(n+1)} = 0$

$$\frac{U_{i+1/2,j}^{(n)} - U_{i-1/2,j}^{(n)}}{\Delta x_i} + \frac{V_{i,j+1/2}^{(n)} - V_{i,j-1/2}^{(n)}}{\Delta y_j} +$$
$$+ \Delta t \left[\frac{1}{\Delta x_i}\left(\frac{1}{\Delta x_{i+1/2}} + \frac{1}{\Delta x_{i-1/2}}\right) + \frac{1}{\Delta y_j}\left(\frac{1}{\Delta y_{j+1/2}} + \frac{1}{\Delta y_{j-1/2}}\right)\right] \Delta\varphi_{i,j} = 0 \tag{5.14}$$

or

$$\Delta\varphi_{i,j} = -\frac{S}{\left(\frac{\partial S}{\partial \varphi}\right)} \tag{5.15}$$

with

$$S = D_{i,j}^{(n)}$$

$$\frac{\partial S}{\partial \varphi} = \Delta t[\frac{1}{\Delta x_i}(\frac{2}{\Delta x_{i+1}+\Delta x_i}+\frac{2}{\Delta x_i+\Delta x_{i-1}})+\frac{1}{\Delta y_j}(\frac{2}{\Delta y_{j+1}+\Delta y_j}+\frac{2}{\Delta y_j+\Delta y_{j-1}})] \quad (5.16)$$

Convergence of (5.15) (which is a variant of the Newton-Raphson relaxation technique) can be accelerated if $D_{i,j}^{(n)}$ is multiplied by an over-relaxation factor ω such that $1 \leq \omega \leq 2$. A value of ω which is often optimum is 1.7 ; an unstable iteration results if ω exceeds 2 (Nichols et al. , 1980). After the pressure correction is found from (5.15), neighbor velocities are updated using (5.13.a)–(5.13.d) ; the pressure correction is always computed using the most up-to-date velocities.

Pressure iteration in a surface cell

In cells containing a free-surface, the dynamic free-surface boundary condition replaces the continuity equation. If the fluid is treated as inviscid, the surface cell pressure is given by

$$\varphi_{i,j} = (1-\eta_s)\varphi_f + \eta_s\varphi_s$$

where φ_f is the pressure in the nearest neighbor full cell, and $\eta_s = \frac{d_c}{d_{fs}}$ is the ratio of the distance between the centers of the full and surface cells, and the distance between the free-surface and the center of the interpolation (full) cell. Equation (5.15) can still be used to compute the pressure correction for a surface cell, provided S is replaced by

$$S = (1-\eta_s)\varphi_f - \eta_s\varphi_s - \varphi_{i,j} \quad (5.17)$$

One complete iteration consists of correcting the pressures and velocities in all fluid-occupied cells. Because the pressure in any one cell affects the pressures in all other cells, the mesh must be swept several times until convergence is reached.

Convergence is attained when the maximum absolute value of S in all cells falls below some small tolerance value $\hat{\epsilon}$, specified by the user. Hirt & Cook (1972) suggest $\hat{\epsilon} = 10^{-2}U/\Delta s$, where $U/\Delta s$ is a typical velocity to length ratio. Choosing $\hat{\epsilon}$ too small results in a high number of pressure iterations per cycle, whereas if $\hat{\epsilon}$ is too large the flow becomes "numerically compressible" and the residual divergences may cause pressure-velocity instabilities. In the present work, it was found that choosing $\hat{\epsilon} = 10^{-3}U/\Delta s$ is often a good choice. The values of Δt and $\hat{\epsilon}$ should be chosen in such a way that the average number of pressure iterations per cycle falls between 10 and 25.

If the viscosity is high, which means that $\frac{U\Delta s}{\nu}$ is small (say, below 10), the inviscid dynamic boundary condition is no longer correct. In this case φ_s must be replaced by $\varphi_s+2\nu(\partial U_n/\partial n)^n_{i,j}$ in (5.17) . In 2D-HYDROTUR, the user specifies the type of dynamic free-surface boundary conditions by means of an indicator. If the viscous normal free-surface boundary condition is used, the tangential stress boundary condition appropriate for a viscous fluid is also used. With this option, it is possible to simulate flows with small and high Reynolds numbers. It is also possible to investigate a particular problem with both types of free-surface boundary conditions and compare the results.

5.4 Approximations for the K and ϵ equations.

The discrete approximations for the K and ϵ equations must minimize numerical diffusion, and reduce the stiffness inherent to the production and dissipation terms. In addition, the solution herein proposed is based on the following two rules :

- The time variation, convection and diffusion terms are globally treated using the modified FTCS/upwind scheme.

- A certain degree of implicitness is included in the production and dissipation terms.

In the present solution algorithm, the time-advanced velocity field ($U^{n+1}_{i,j}$, $V^{n+1}_{i,j}$) is already available when the K-ϵ equations are updated. Therefore, the production term $\mathcal{P}$ and the convection velocity can be computed in both $n+1$ and n time levels. Hence, it is possible to improve time-centering and at the same time retain the basic explicit character of the scheme. In the code development stage, four different schemes were tested :

Scheme 1

$$\frac{K^{n+1}_{i,j}-K^n_{i,j}}{\Delta t} = -FKX-FKY+VISK+\mathcal{P}^{n+1}_{i,j}-\epsilon^n_{i,j}$$

$$\frac{\epsilon^{n+1}_{i,j}-\epsilon^n_{i,j}}{\Delta t} = -FEX-FEY+VISE+C_{1\epsilon}(\frac{\epsilon}{K})^n_{i,j}\mathcal{P}^{n+1}_{i,j}-C_{2\epsilon}(\frac{\epsilon^2}{K})^n_{i,j}$$

Scheme 2

$$\frac{K^{n+1}_{i,j}-K^n_{i,j}}{\Delta t} = -FKX-FKY+VISK+\mathcal{P}^{n+1}_{i,j}-\epsilon^n_{i,j}$$

$$\frac{\epsilon^{n+1}_{i,j}-\epsilon^n_{i,j}}{\Delta t} = -FEX-FEY+VISE+C_{1\epsilon}(\frac{\epsilon}{K})^{n+1}_{i,j}\mathcal{P}^{n+1}_{i,j}-C_{2\epsilon}(\frac{\epsilon}{K})^n_{i,j}\epsilon^{n+1}_{i,j}$$

Scheme 3

$$\frac{\epsilon_{i,j}^{n+1}-\epsilon_{i,j}^{n}}{\Delta t}=-FEX-FEY+VISE+C_{1\epsilon}\left(\frac{\mathcal{P}_{i,j}^{n+1}}{K_{i,j}^{n}}\right)\epsilon_{i,j}^{n+1}-C_{2\epsilon}\left(\frac{\epsilon}{K}\right)_{i,j}^{n}\epsilon_{i,j}^{n+1}$$

$$\frac{K_{i,j}^{n+1}-K_{i,j}^{n}}{\Delta t}=-FKX-FKY+VISK+\mathcal{P}_{i,j}^{n+1}-\epsilon_{i,j}^{n+1}$$

Scheme 4

$$\frac{\epsilon_{i,j}^{n+1}-\epsilon_{i,j}^{n}}{\Delta t}=-FEX-FEY+VISE+C_{1\epsilon}\left(\frac{\epsilon_{i,j}^{n}}{K_{i,j}^{n}}\right)\mathcal{P}_{i,j}^{n+1}-C_{2\epsilon}\left(\frac{\epsilon}{K}\right)_{i,j}^{n}\epsilon_{i,j}^{n+1}$$

$$\frac{K_{i,j}^{n+1}-K_{i,j}^{n}}{\Delta t}=-FKX-FKY+VISK+\frac{1}{2}(\mathcal{P}_{i,j}^{n+1}+\mathcal{P}_{i,j}^{n}-\epsilon_{i,j}^{n+1}-\epsilon_{i,j}^{n})$$

In these equations, the following definitions have been introduced

$$\mathcal{P}_{i,j}^{n+1}=(\nu_t)_{i,j}^{n}[2DUDX^2+2DVDY^2+(DUDY+DVDX)^2]$$

$$FKX=\frac{U_{i,j}^{n+1}}{\Delta x_\alpha}[\Delta x_r DKDL+\Delta x_l DKDR+\gamma sign(U_{i,j}^{n+1})(\Delta x_r DKDL-\Delta x_l DKDR)]$$

$$FKY=\frac{V_{i,j}^{n+1}}{\Delta y_\alpha}[\Delta y_t DKDB+\Delta y_b DKDT+\gamma sign(V_{i,j}^{n+1})(\Delta y_t DKDB-\Delta y_b DKDT)]$$

$$FEX=\frac{U_{i,j}^{n+1}}{\Delta x_\alpha}[\Delta x_r DEDL+\Delta x_l DEDR+\gamma sign(U_{i,j}^{n+1})(\Delta x_r DEDL-\Delta x_l DEDR)]$$

$$FEY=\frac{V_{i,j}^{n+1}}{\Delta y_\alpha}[\Delta y_t DEDB+\Delta y_b DEDT+\gamma sign(V_{i,j}^{n+1})(\Delta y_t DEDB-\Delta y_b DEDT)]$$

$$VISK=\Delta_x^1\left[\left(\frac{\nu_t}{\sigma_k}+\nu\right)\Delta_x^1 K\right]_{i,j}^{n}+\Delta_y^1\left[\left(\frac{\nu_t}{\sigma_k}+\nu\right)\Delta_y^1 K\right]_{i,j}^{n}+$$

$$+\frac{1}{2}(1-\gamma)\Delta t(U^2)_{i,j}^{n+1}\Delta_{xx}K_{i,j}^{n}+\frac{1}{2}(1-\gamma)\Delta t(V^2)_{i,j}^{n+1}\Delta_{yy}K_{i,j}^{n}$$

$$VISE=\Delta_x^1\left[\left(\frac{\nu_t}{\sigma_\epsilon}+\nu\right)\Delta_x^1\epsilon\right]_{i,j}^{n}+\Delta_y^1\left[\left(\frac{\nu_t}{\sigma_\epsilon}+\nu\right)\Delta_y^1\epsilon\right]_{i,j}^{n}+$$

$$+\frac{1}{2}(1-\gamma)\Delta t(U^2)_{i,j}^{n+1}\Delta_{xx}\epsilon_{i,j}^{n}+\frac{1}{2}(1-\gamma)\Delta t(V^2)_{i,j}^{n+1}\Delta_{yy}\epsilon_{i,j}^{n}$$

The quantities appearing in these equations have an obvious meaning, *viz.*

$$DUDX=\Delta_x^1 U_{i,j}^{n+1} \qquad DUDY=\Delta_y^0 U_{i,j}^{n+1}$$

$$DVDY=\Delta_y^1 V_{i,j}^{n+1} \qquad DVDX=\Delta_y^0 V_{i,j}^{n+1}$$

$$DKDL=\Delta_x^- K_{i,j}^{n} \qquad DEDL=\Delta_x^-\epsilon_{i,j}^{n}$$

$$DKDR=\Delta_x^+ K_{i,j}^{n} \qquad DEDR=\Delta_x^+\epsilon_{i,j}^{n}$$

$$DKDB = \Delta_y^- K_{i,j}^n \qquad DEDB = \Delta_y^- \epsilon_{i,j}^n$$

$$DKDT = \Delta_y^+ K_{i,j}^n \qquad DEDT = \Delta_y^+ \epsilon_{i,j}^n$$

The space increments Δx_r , Δx_l , Δy_t and Δy_b , are the distances between the center of cell ij and the center of its right, left, top and bottom neighbors, respectively. The space increments Δx_α and Δy_α are defined by the expressions

$$\Delta x_\alpha = \Delta x_r + \Delta x_l + \gamma sign(\Delta x_r - \Delta x_l)$$
$$\Delta y_\alpha = \Delta y_t + \Delta y_l + \gamma sign(\Delta x_t - \Delta x_l)$$

The convection-diffusion part of the K-ϵ equations is approximated by means of hybrid modified FTCS/upwind differencing, with a weighting factor γ that controls the amount of each one. $\gamma = 1$ corresponds to upwind differencing, whereas $\gamma = 0$ gives the modified FTCS scheme. The approximations to the spatial derivatives are also weighted by the spatial increments, in order to preserve formal accuracy in the variable mesh. As was done with the approximations for the momentum equations, the eddy-viscosity is multiplied by the damping function $\frac{\nu_t}{\beta\nu}[1 - exp(-\beta\frac{\nu}{\nu_t})]$ (see § 5.2).

The differences between schemes 1—4 are the discretization of the production and dissipation terms, the order in which the K and ϵ equations are advanced in time, and the way in which information on the time level $n+1$ is incorporated in the solution.

Scheme 1 is the most straightforward alternative. The order in which the two equations are advanced is irrelevant. This scheme gave good results in some test problems, but in certain cases the predictions were poor. For instance, the computed profiles for K, ϵ and ν_t in a logarithmic boundary layer showed relative errors of up to 15% with respect to their theoretical values.

In scheme 2, the K equation is solved first, and then $K_{i,j}^{n+1}$ is used in the production/dissipation terms in the discretized ϵ equation. The dissipation term in the ϵ equation is linearized, and the production term is treated implicitly. This scheme was found to be superior to scheme 1; it produced better simulations in the closed-cavity and logarithmic-layer test problems.

In scheme 3, the ϵ equation is updated first, and the $(Production - Dissipation)_\epsilon$ term is treated in a semi-implicit way. The interest of this alternative is the appearance of $\mathcal{P}^{n+1} - \epsilon^{n+1}$ in the K equation. The fact that $\mathcal{P}$ and ϵ are in the advanced time level improves the performance of the model in situations where the $\mathcal{P} - \epsilon$ term dominates the other terms.

Scheme 4 is based on the same ideas of scheme 3, but with two variants. First, only the dissipation term in the ϵ equation is treated in a semi-implicit way. Second the $\mathcal{P} - \epsilon$ term in the K equation is centered in time with $(K_{i,j}^{n+1} - K_{i,j}^n)/\Delta t$. This arrangement is based on the following assumptions :

- In situations where $(\mathcal{P}-\epsilon)_\iota$ is small, it is better to keep a degree of implicitness only in the dissipation term.

- The time centering of $\frac{K_{i,j}^{n+1}-K_{i,j}^{n}}{\Delta t}$ and $\mathcal{P}-\epsilon$ improves the performance of the K equation.

After comparing the predictions of all these schemes in the isotropic turbulence decay, logarithmic boundary layer, and turbulent cavity flow test problems (§ 5.11), scheme 4 was adopted in the final version of the code. Accordingly, the final discretized K-ϵ equations are

$$\epsilon_{i,j}^{n+1}=\frac{1}{\frac{1}{\Delta t}+C_{2\epsilon}(\frac{\epsilon}{K})_{i,j}^{n}}(\frac{\epsilon_{i,j}^{n}}{\Delta t}-FEX-FEY+VISE+C_{1\epsilon}(\frac{\epsilon}{K})_{i,j}^{n}\mathcal{P}_{i,j}^{n+1}) \tag{5.18}$$

$$K_{i,j}^{n+1}=K_{i,j}^{n}+\Delta t(-FKX-FKY+VISK+\tfrac{1}{2}[(\mathcal{P}-\epsilon)_{i,j}^{n+1}+(\mathcal{P}-\epsilon)_{i,j}^{n}] \tag{5.19}$$

If both production and dissipation terms in (5.18) were treated in a semi-implicit way, the denominator in (5.18) would have been

$$\frac{1}{\Delta t}+\frac{1}{K_{i,j}^{n}}(C_{2\epsilon}\epsilon_{i,j}^{n}-C_{1\epsilon}\mathcal{P}_{i,j}^{n+1})$$

which is smaller than $(1/\Delta t)+C_{2\epsilon}(\epsilon/K)_{i,j}^{n}$. This aggravates the inherent stiffness of the ϵ equation.

5.5 Updating the fluid configuration.

After the velocity, pressure and K-ϵ (if $\iota=1$) fields are advanced to time level $n+1$, the fluid configuration is updated. Once the time-advanced fluid configuration is found, appropriate boundary conditions are imposed and the cycle is restarted.

5.5.1 Algorithm for the convection of F.

The VOF function F is governed by equation (5.4). Because F is discontinuous, its convective fluxes must be computed in a special way that preserves sharp interface definition, conserves the total fluid volume, and leads to a numerically stable algorithm. Combining (5.4) with the continuity equation, the equation for the time evolution of F can be put in divergence or conservative form

$$\frac{\partial F}{\partial t}+\frac{\partial(FU)}{\partial x}+\frac{\partial(FV)}{\partial y}=0 \tag{5.20}$$

When (5.20) is integrated over a computational cell, the changes in F in that cell reduce to fluxes of F across the cell faces. Following Nichols & Hirt (1975), the donor-acceptor flux approximation is used. The flux of volume crossing the cell face is $U_x = |U|\Delta t$, where U is the normal velocity at the face. The amount of F that is actually fluxed across the cell face is

$$\Delta F = \Delta y_j min\{F_{AD}|U_x|+CF, F_D \Delta x_D\} \tag{5.21}$$

with

$$CF = max\{(1-F_{AD})|U_x|-(1-F_D)\Delta x_D, 0\} \tag{5.22}$$

In the same way, the flux of volume crossing the top boundary of a cell ij is $V_y = |V|\Delta t$, and the amount of F actually fluxed through the top cell face is

$$\Delta F = \Delta x_i min\{F_{AD}|V_y|+CF, F_D \Delta y_D\} \tag{5.23}$$

with

$$CF = max\{(1-F_{AD})|V_y|-(1-F_D)\Delta y_D, 0\} \tag{5.24}$$

In these expressions, the subscript A denotes the acceptor cell, the subscript D the donor cell, and the double subscript AD is either A or D depending on the orientation of the free-surface relative to the direction of the flow. The *min* feature in these equations prevents the fluxing of more F from the donor cell than it has to give, while the *max* feature accounts for an additional F flux (CF) if the amount of void ($1-F$) to be fluxed exceeds the amount of void available in the donor cell. The *min* and *max* tests prevent the unbounded growth of potential instability when the acceptor cell is used to compute the flux (*i.e.* when $AD = A$).

Following Nichols ét al. (1980), the rules for choosing $AD = A$ or $AD = D$ are the following. If the free-surface is convected mostly normal to itself, the acceptor cell F-value is used, otherwise the donor cell F-value is used. If the acceptor cell is empty, or if the cell upstream of the donor cell is empty, the acceptor cell is used to determine the flux, regardless of the orientation of the free-surface. In this way, the sharpness of the interface is preserved without incorrect steepening of surface waves.

The value ΔF computed by the above method is subtracted from the donor cell and added to the acceptor cell. This process is repeated for all cell boundaries in the mesh, and the resulting F values define the new fluid configuration. Occasionally, the time-advanced F values may have values slightly less than zero, or slightly greater than 1. Hence, after the F-convection calculation has been completed, the mesh is swept to reset values of F less than zero back to zero, and values of F greater than one back to one. Accumulated changes in F are recorded during a computation; they may be printed at any time and are available in the program's run-time message (cycle information). These tiny volume falsifications are due to small errors in mass conservation and numerical diffusion in the F-convection algorithm. They can be reduced by either specifying a smaller tolerance for the residual divergence in the pressure-velocity iteration, or by increasing the mesh resolution and lowering the Courant number.

A free-surface cell i,j is defined as a cell with $0<F<1$ and having at least one neighboring cell $(i\pm1,j)$ or $(i,j\pm1)$ that is empty (*i.e.* contains a zero value of F). However, F values cannot be tested against exact numbers because of roundoff errors. Instead, a cell is defined to be empty when F is less than ϵ_f and full when F is greater than $1-\epsilon_f$, where ϵ_f is typically 10^{-6}. Accumulated volume errors after hundreds of cycles are typically a fraction of a per cent of the total F volume. This indicates that the F-convection algorithm preserves free-surface sharpness and has good conservation properties.

5.5.2 Determining interfaces within a cell.

For the accurate application of the normal-stress free-surface boundary condition, it is necessary to determine the location of the free-surface within the surface cell. In addition, it may be necessary to know the local free-surface curvature (*e.g.* if surface tension effects are to be included). Following Nichols et al. (1980), a line segment is fitted into each interfacial cell by first determining its slope, and then moving the segment across the cell to a position that gives the correct amount of fluid-occupied volume in the cell. A brief survey of methods for determining free-surface location and curvature in two dimensions using the F-function technique can be found in Poo & Ashgriz (1989).

The basic problem is illustrated in figure 5.4. To determine the slope, it is assumed that the boundary can be represented either as a single-valued function $Y(x)$ or $X(y)$, depending on its orientation. If the boundary is representable as $Y(x)$, $\frac{dY}{dx}$ may be computed in the following way

$$Y_i = F_{i,j-1}\Delta y_{i-1} + F_{i,j}\Delta y_i + F_{i,j+1}\Delta y_{j+1}$$

$$\left(\frac{dY}{dx}\right)_i = \left(\frac{Y_{i+1}-Y_i}{\Delta x_{i+1/2}}\Delta x_{i-1/2} + \frac{Y_i - Y_{i-1}}{\Delta x_{i-1/2}}\Delta x_{i+1/2}\right)\frac{1}{\Delta x_{i+1/2}+\Delta x_{i-1/2}}$$

In the same way, if the boundary is representable as $X(y)$, $\frac{dX}{dy}$ may be found using the following expressions

$$X_j = F_{i-1,j}\Delta x_{i-1} + F_{i,j}\Delta x_i + F_{i+1,j}\Delta x_{i+1}$$

$$(\frac{dX}{dy})_j = (\frac{X_{j+1}-X_j}{\Delta y_{j+1/2}}\Delta y_{j-1/2} + \frac{X_j - X_{j-1}}{\Delta y_{j-1/2}}\Delta y_{j+1/2})\frac{1}{\Delta y_{j+1/2}+\Delta y_{j-1/2}}$$

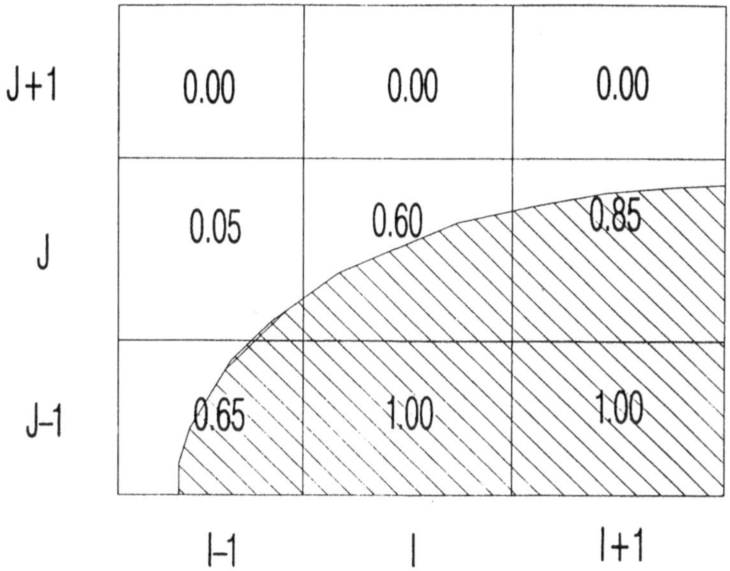

Figure 5.4 : *Determining interface position within a surface cell.*

If $|dY/dx|_i < |dX/dy|_j$, the boundary is more nearly horizontal, otherwise it is more nearly vertical. The derivative with the smallest magnitude gives the best approximation to the local slope. The pair of values $(dY/dx)_i$ and $(dX/dy)_j$ may be used to find the side in which the fluid lies. For instance, if $|dY/dx|$ is smallest and the surface is more nearly horizontal than vertical, a negative $(dX/dy)_j$ implies that the fluid lies below the free-surface. In that case, the cell $i,j-1$ would be identified as the pressure interpolation cell. Once the boundary slope and the side occupied by the fluid have been determined, a line is fitted through the cell with the correct amount of volume $F_{i,j}\Delta x_i \Delta y_j$ lying on the fluid side.

5.6 Velocity boundary conditions.

5.6.1 Mesh boundaries.

At the mesh boundaries, several types of boundary conditions may be imposed using the layer of fictitious cells surrounding the mesh. The built-in types of boundary conditions are :

- Free-slip boundary conditions.
- No-slip boundary conditions.

- Continuative (outflow) boundary conditions.

- Constant pressure boundary conditions.

- Periodic boundary conditions.

- Law-of-the-wall (partial slip) boundary conditions.

Consider a left boundary, illustrated in figure 5.5 . If this is a rigid free-slip wall, the normal velocity and the gradient of the tangential velocity are both set to zero

$$U_{1,j}^{n+1}=0 \qquad V_{1,j}^{n+1}=V_{2,j}^{n+1} \qquad \varphi_{1,j}=\varphi_{2,j} \qquad F_{1,j}^{n+1}=F_{2,j}^{n+1}$$

for all j. If the left boundary is a rigid no-slip wall, both normal and tangential velocity components are set to zero, and the appropriate boundary conditions are

$$U_{1,j}^{n+1}=0 \qquad V_{1,j}^{n+1}=-V_{2,j}^{n+1}\frac{\Delta x_1}{\Delta x_2} \qquad \varphi_{1,j}=\varphi_{2,j} \qquad F_{1,j}^{n+1}=F_{2,j}^{n+1}$$

for all j. For a continuative or outflow boundary, a prescription is needed that permits the fluid to flow out of the domain with a minimum of upstream influence. Following Nichols et al. (1980), the continuative boundary conditions imposed at the left wall are

$$U_{1,j}^{n+1}=U_{2,j}^{n+1} \qquad V_{1,j}^{n+1}=V_{2,j}^{n+1} \qquad \varphi_{1,j}=\varphi_{2,j} \qquad F_{1,j}^{n+1}=F_{2,j}^{n+1}$$

for all j. For periodic boundary conditions in the x-direction, the left and right boundaries must be set to reflect the periodicity. The period length is chosen equal to the distance from the left wall to the left of the last interior cell in the mesh at the right hand side. Hence the I = IMAX and I = IMAX−1 columns of cells are reserved on the right side of the mesh for setting periodic boundary conditions, which are

$$U_{1,j}^{n+1}=U_{imax-2,j}^{n+1} \qquad V_{1,j}^{n+1}=V_{imax-2,j}^{n+1} \qquad F_{1,j}^{n+1}=F_{imax-2,j}^{n+1}$$

on the left, and

$$U_{imax-1,j}^{n+1}=U_{2,j}^{n+1} \qquad V_{imax-1,j}^{n+1}=V_{2,j}^{n+1} \qquad \varphi_{imax-1,j}=\varphi_{2,j}$$

$$V_{imax,j}^{n+1}=V_{3,j}^{n+1} \qquad F_{imax,j}^{n+1}=F_{3,j}^{n+1} \qquad F_{imax-1,j}^{n+1}=F_{2,j}^{n+1}$$

on the right, for all j.

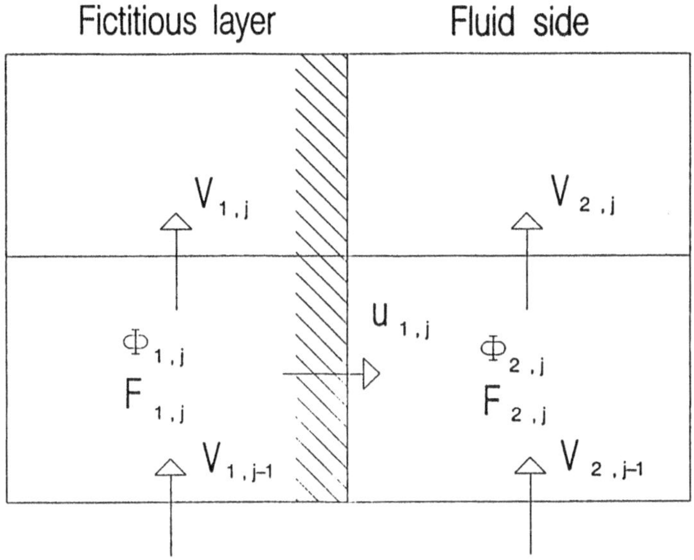

Figure 5.5 : *Velocity boundary conditions near a left wall.*

A constant pressure boundary condition at the left wall is set by keeping the pressure in column $I=2$ constant and otherwise treating the boundary as continuative.

When the law-of-the-wall is employed in turbulent flow simulations, the normal gradient of the tangential velocity is specified according to the formulae

$$\frac{V_{2,J}}{u_*} = \frac{1}{\kappa} ln(E \frac{u_* \Delta x_2}{2\nu})$$

$$\left|\frac{\partial V}{\partial x}\right|_{\Sigma} = \frac{2u_*}{\kappa \Delta x_2}$$

$$V_{1,J}^{n+1} = V_{2,J}^{n+1} - sign(V_{2,J}) \left|\frac{\partial V}{\partial x}\right|_{\Sigma} \frac{\Delta x_1 + \Delta x_2}{2}$$

The computation of u_* is discussed in § 5.7 .

In the free-slip, no-slip, periodic and partial-slip/law-of-the-wall cases, the boundary conditions are imposed after applying the momentum equations and after each pass through the pressure iteration. In the continuative and constant pressure cases, the above conditions are imposed after applying the momentum equations, but not during the pressure iteration.

Boundary conditions similar to the ones for the left mesh boundary are used at the right, top and bottom boundaries, reversing the roles of U and V as normal and tangential velocities when appropriate. Any of the above types of boundary conditions can be automatically imposed by using a code of input numbers (WL, WR, WT, and WB), whose values can be : 1 for a rigid free-slip boundary; 2 for a rigid no-slip boundary; 3 for a continuative (outflow) boundary; 4 for a periodic boundary; and 5 for a constant pressure boundary. Law-of-the-wall boundary conditions are automatically selected, regardless of indicator numbers, when the turbulence model indicator (ITL10) is set to one (indicating a turbulent flow simulation with the K-ϵ model), unless special boundary conditions that override the default settings are programmed.

Special inflow or outflow boundary conditions at mesh boundaries or at locations within the mesh can be coded by hand in a special location in the program reserved for that purpose.

5.6.2 Free-surface boundaries.

Free-surface boundary conditions are imposed in several ways in the solution algorithm. The kinematic boundary condition is automatically satisfied in the F-advection step. The normal stress dynamic boundary condition is imposed during the implicit pressure evaluation. It remains to satisfy the tangential stress boundary condition.

The tangential stress condition is approximated by specifying velocities immediately outside the free-surface, where these values are needed in the finite-difference equations for points at the surface cells. The specifications herein proposed are identical to those of previous codes, but incorporating both inviscid and viscous cases (§ 4.7.3). The selection between these two options is done by means of a user specified indicator.

If the surface cell has only one empty neighbor cell, the velocity in the face separating them is chosen so that $D_{i,j}=0$ at the surface cell. When there are two or more empty neighbor cells, the individual contributions to the divergence $\partial U/\partial x$ and $\partial V/\partial y$ are separately set to zero. The $D_{i,j}=0$ condition is supplemented by prescribing tangent velocities in empty cells adjacent to partial fluid cells. There are four possible cases, as represented in figure 5.6 ; the surface cell as indices i and j , and the velocity to be specified (represented by a black dot) is to be found as a function of the other three velocities (black squares). If the fluid is treated as inviscid, zero values for $\partial U/\partial y$ or $\partial V/\partial x$ are used to find the unknown velocity. This is a version of the simplified condition $\partial U_t/\partial n=0$ (§ 4.7.3). If the fluid is treated as viscous, the unknown velocity is found using the condition $\partial U/\partial y+\partial V/\partial x=0$, which is a simplified version of $\partial U_n/\partial t+\partial U_t/\partial n=0$ (§ 4.7.3).

5.6.3 Internal-obstacle boundaries.

Internal obstacles are defined by flagging all mesh cells that are to be blocked. In order to reduce the number of arithmetic operations, the pressure relaxation factor $1/(\partial S/\partial\varphi)$ is stored in an array BETA(I,J) in the code. Because the relaxation factor must be positive, negative values of this variable can be used as a flag for identifying obstacle cells. A convenient way to do this is to assign a value of BETA(I,J) = −1 to all obstacle cells. The flag values for BETA must be programmed in the code for each application, in a section reserved for that purpose.

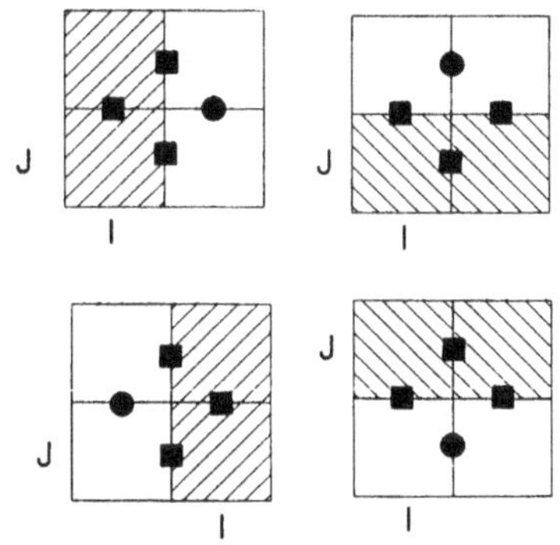

Figure 5.6 : *Tangential stress boundary condition.*

Velocities and pressures are not calculated in obstacle cells (§§ 5.2 and 5.3). All velocity components on faces of obstacle cells are automatically set to zero (figure 5.7). In the computer program, three types of boundary conditions are automatically provided in cells adjacent to obstacle cells: no-slip, free-slip, and law-of-the-wall boundary conditions.

No-slip conditions are specified by setting all velocities equal to zero inside obstacle cells. Because the tangential velocities are zero at locations shifted into the obstacles one-half of a cell width from the actual boundary location, the tangential velocity conditions are only first-order accurate. This problem is inherent to the structure of the MAC mesh (§ 4.2).

Free-slip boundary conditions may be imposed by setting $\partial U/\partial y$ or $\partial V/\partial x$ equal to zero (in the same way as for a free-surface). For instance, in figure 5.7 $V_{i,j+3/2} = V_{i,j+1/2}$.

When the turbulence model is used, law-of-the-wall prescriptions identical to those for regular mesh boundaries are imposed. In this case, general obstacle configurations are not automatically provided. Instead, the bottom boundary is assumed to be of the form $h_b = h_b(x)$, and the cells are blocked by defining DO-LOOP index bounds as JBB(I)+1 and JMAX−1, instead of 2 and JMAX−1 (figure 5.8). This simplifies the internal coding of the DO-LOOPS, favors vectorization, and is adequate for the purposes of the present work. More general obstacle configurations (for instance, a submerged pipeline) can still be programmed by hand.

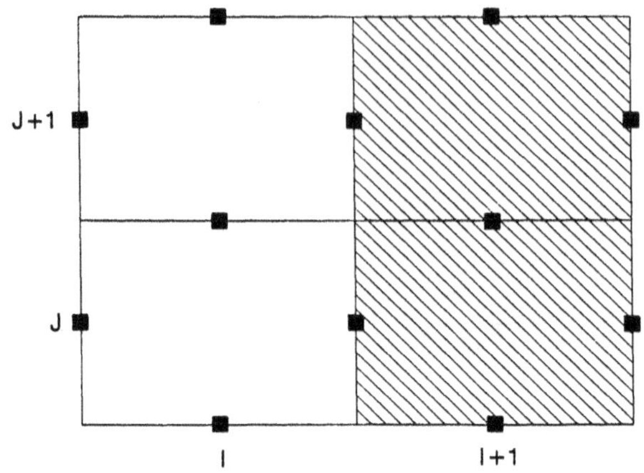

Figure 5.7 . *Obstacle boundary conditions.*

5.7 Boundary conditions for the K and ϵ equations.

In 2D-HYDROTUR, the following types of boundary conditions for K and ϵ are automatically provided : law-of-the-wall; outflow (continuative); and free-surface.

Inflow boundary conditions, which are problem-dependent, can be coded by hand in a special section of the program indicated for that purpose.

In cells adjacent to a wall boundary, the local state of turbulence is expressed in terms of the friction velocity u_* . In the computer program, u_* is calculated using the logarithmic law-of-the-wall for a smooth boundary $U^+ = (1/\kappa)ln(Ey^+)$, with $U^+ = U_t/u_*$, $y^+ = u_* y/\nu$ and $U_t =$ *tangent velocity in the cell next to the wall*. The logarithmic law-of-the-wall is an implicit equation in u_* . This implicit equation is solved by a Newton-Raphson iteration

$$X^{(n+1)} = X^{(n)} + \frac{\kappa - X^{(n)}(A + ln(X^{(n)}))}{1 + A + ln(X^{(n)})} \tag{5.25}$$

with $A = ln(EU_t y/\nu)$ and $X = u_*/U_t$. Typical values of X range from 9×10^{-2} to 4×10^{-2} . The convergence of (5.25) is extremely fast. Two iterations are enough to determine X with a relative error of 0.05% . Therefore, convergence tests need not be performed, and this allows the full vectorization of all DO-loops where the calculation of u_* is required. If $y^+ < 40$ (corresponding to the buffer and linear layers), viscous damping effects become important, and the logarithmic law-of-the-wall is not valid. In the present work, u_* is found by a combination of the logarithmic law-of-the-wall and the linear law $U^+ = y^+$

$$u_* = max\{X|U_t|, \sqrt{|U_t|\nu/y}\} \tag{5.26}$$

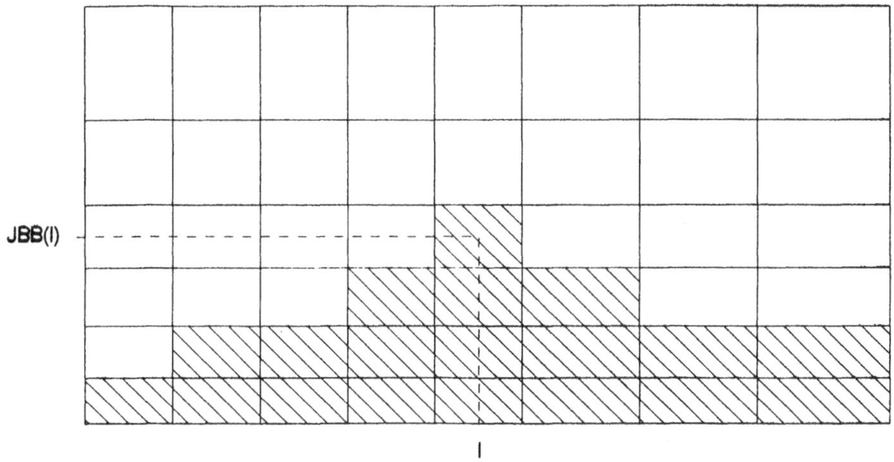

Figure 5.8 : *Configuration of the bottom boundary for the turbulence model.*

This combination is slightly inconsistent in the buffer-layer, $3 < y^+ < 40$. However, the more complicated expressions that have been proposed for the velocity profile in the buffer-layer are not entirely satisfactory (Hinze, 1975), and make the calculation of u_* less efficient.

The boundary conditions for K and ϵ near the wall are computed using the following expressions (§ 2.10.1)

$$K = u_*^2/(C_d)^{1/2} \tag{5.27}$$

$$\epsilon = u_*^3/[\kappa y(1 - exp(-\frac{u_* y}{26\nu}))] \tag{5.28}$$

The modification of (5.28) with respect to the usual expression $\epsilon = u_*^3/\kappa y$ gives the following equation for the eddy-viscosity

$$\nu_t = \kappa u_* y[1 - exp(-\frac{u_* y}{26\nu})] \tag{5.29}$$

This is the van-Driest law (Launder & Spalding, 1972; Hinze, 1975), which accounts for direct viscous dissipation when the turbulence intensity is low. The van-Driest damping function is confirmed by experiments in many wall flows (Launder & Spalding, 1972). The velocity profile consistent with the van-Driest law is (Hinze, 1975) :

$$U^+ = 2\int_0^{y^+} \frac{dz}{1 + \{1 + 4\kappa^2 z^2 [1 - exp(-z/26)]^2\}^{1/2}} \tag{5.30}$$

However, deviations of (5.30) from the linear-log profile combination in the buffer-layer are small.

For an outflow boundary, homogeneous Neumann conditions for K and ϵ are imposed. For a right mesh boundary, this gives

$$K^{n+1}_{imax-1,j} = K^{n+1}_{imax-2,j} \qquad \epsilon^{n+1}_{imax-1,j} = \epsilon^{n+1}_{imax-2,j}$$

Similar expression are employed for other outflow boundaries.

The free-surface boundary conditions for K and ϵ are the most difficult to specify, for there is no universally accepted solution. For the purposes of the present work, neither Rodi ´s (1980) proposal nor variants of the law-of-the-wall are usable. These alternatives assume that *i)* the free-surface has small curvature, and *ii)* the wind stress is the primary source of turbulence intensity.

In view of the flows herein considered (chapter 3) and of the assumed restrictions of the present formulation, the following criteria are proposed :

a) All relevant transport processes occur inside the water body and are driven by internal velocity shear (mainly caused by the folding of the free-surface onto itself during the breaking process).

b) The influence of air in the problem is neglected.

The limitations of such criteria for the simulation of real breaking flows were discussed in §§ 2.10 and 3.2 . In order to derive a simple free-surface boundary condition for K and ϵ, a) and b) are supplemented by the following assumptions :

c) Turbulence does not diffuse across the free-surface.

d) Cross-stream convection is neglected at the free-surface.

The simplified K and ϵ equations resulting from c) and d) are

$$\frac{\partial K}{\partial t} + U_s \frac{\partial K}{\partial s} = \frac{\partial}{\partial s}[(\frac{\nu_t}{\sigma_k} + \nu)\frac{\partial K}{\partial s}] + \mathcal{P} - \epsilon \tag{5.31}$$

$$\frac{\partial \epsilon}{\partial t} + U_s \frac{\partial \epsilon}{\partial s} = \frac{\partial}{\partial s}[(\frac{\nu_t}{\sigma_\epsilon} + \nu)\frac{\partial \epsilon}{\partial s}] + C_{1\epsilon}\frac{\epsilon}{K}\mathcal{P} - C_{2\epsilon}\frac{\epsilon^2}{K} \tag{5.32}$$

In (5.31)–(5.32), U_s is the streamwise velocity, and $\partial/\partial s$ denotes the directional derivative in the streamwise direction. Free-surface boundary conditions for K and ϵ are imposed by setting K and ϵ in empty cells adjacent to surface cells. Starting from a)–d) and (5.31)–(5.32) a further simplification is introduced :

e) If ij is a surface cell and its neighbor $i\pm1,j$ is empty, then $U_s \simeq U$ and the free-surface is assumed horizontal; if the neighbor $i,j\pm1$ is empty, $U_s \simeq V$ and the free-surface is assumed vertical.

This simplification is very similar to the one used in the formulation of the free-surface boundary conditions for velocity (tangential stress condition) in the momentum equations.

Using a)–e) , equations (5.31) and (5.32) result from the full K and ϵ equations if $\partial K/\partial x = \partial\epsilon/\partial x = 0$ is imposed for a nearly vertical surface, and $\partial K/\partial y = \partial\epsilon/\partial y = 0$ is imposed for a nearly horizontal free-surface. When the surface cell has two or more adjacent empty cells, a combination of both conditions is used.

Thus, the simplest solution that is compatible with a)–e) is the prescription of homogeneous Neumann boundary conditions for K and ϵ at the free-surface. It is well known that from the numerical viewpoint this condition has a good stabilizing influence. Physically, this condition means that turbulence does not diffuse across the interface. This is consistent with the hypothesis that the air outside the water body does not influence the internal turbulence-producing mechanisms.

The free-surface boundary conditions for K and ϵ herein proposed are a very crude representation of reality. However, the breaking phenomenon and turbulence in free-surface flows are so complicated that any prescription of boundary conditions is, at present, highly speculative. Nevertheless, it should be mentioned that the use of homogeneous Neumann boundary conditions for turbulence-related variables was recently proposed (in a slightly different context, and with a different theoretical justification) by Launder (1989).

5.8 Initial conditions for the K and ϵ equations.

Ideally, the initial conditions for K and ϵ should be prescribed according to experimental data. However, this type of information is not always available for the flow of interest. Hence, either theoretical or heuristic methods must be used.

The specification of initial conditions for K and ϵ is very seldom discussed in the literature. The typical argument is "since only the steady-state solution is to be found, the initial condition is not very important". This argument is not appropriate for the flows to be considered in this work, because :

a) The flows are unsteady, and the initial condition determines the long-term behavior of the solution considerably.

b) Even for the case of steady flows, a "bad" initial condition increases the computational effort to reach the desired solution.

c) The initial condition cannot be specified in a completely arbitrary manner, because of the singularities in the ϵ equation and the possible decay to negative values.

In the present work, three different methods for generating initial conditions for K and ϵ were used. The first method consisted of guessing typical values for K and ν_t for the problem at hand. Then, constant values for K and ϵ are initially specified in all cells of the computational domain. This uniform initial condition should be combined with the boundary conditions for K and ϵ, in order to avoid as much as possible large destabilizing gradients near the boundaries. This method has the advantage of simplicity. The second method consisted of the application of theoretical profiles for K and ϵ. For instance, a logarithmic profile with a guessed u_* ($K = u_*^2/(C_d)^{1/2} = constant$; $\epsilon = K^{3/2}C_d^{3/4}/\kappa y$) can be used to start a simulation of a turbulent hydraulic jump in a channel, a confined turbulent boundary layer, or a turbulent wave-induced oscillatory boundary layer. This method is often used in the formulation of inflow boundary conditions (Rodi, 1980). This method is also simple, and has a stronger theoretical basis than the first method (constant K and ϵ). However, it is not appropriate for more complex flows, such as recirculating flows past steps or inside cavities. The third method employed is a general-purpose heuristic procedure for the generation of initial distributions for K and ϵ, developed in parallel with the K-ϵ model. The key ideas are :

a) The modeled K and ϵ equations, which define an initial and boundary value problem for K and ϵ, are simplified in such a way that the resulting equations define a boundary value problem for the turbulence variables.

b) The resulting boundary value problem for K and ϵ is solved using the initial K and ϵ boundary conditions.

c) The eddy-viscosity is assumed constant in steps a) and b).

Using a)–c) , the modeled K-ϵ equations become

$$-(U\frac{\partial K}{\partial x}+V\frac{\partial K}{\partial y})+\frac{\nu_i}{\sigma_k}(\frac{\partial^2 K}{\partial x^2}+\frac{\partial^2 K}{\partial y^2})+\mathcal{P}-\epsilon=0 \tag{5.33}$$

$$-(U\frac{\partial \epsilon}{\partial x}+V\frac{\partial \epsilon}{\partial y})+\frac{\nu_i}{\sigma_\epsilon}(\frac{\partial^2 \epsilon}{\partial x^2}+\frac{\partial^2 \epsilon}{\partial y^2})+C_{1\epsilon}\frac{\epsilon}{K}\mathcal{P}-C_{2\epsilon}\frac{\epsilon^2}{K}=0 \tag{5.34}$$

where ν_i is the constant value of the eddy-viscosity used in the simplified boundary value problem defined by (5.33)–(5.34). These equations are steady convection-diffusion equations with source-sink terms. The initializing viscosity ν_i can be chosen equal to the average value of ν_t expected in the flow; $5\times10^{-3} < \nu_i < 5\times10^{-2}$ is a reasonable estimate.

If the initial condition is such that $-\vec{U}\cdot\nabla K$ and $-\vec{V}\cdot\nabla\epsilon$ are small (for instance, in the initiation of a moving hydraulic jump in a channel due to the closing of a gate downstream, with initial law-of-the-wall profiles for U, K and ϵ, and $V=0$) the simplified equations become

$$\frac{\nu_i}{\sigma_K}\left(\frac{\partial^2 K}{\partial x^2}+\frac{\partial^2 K}{\partial y^2}\right)+\mathcal{P}-\epsilon=0 \tag{5.35}$$

$$\frac{\nu_i}{\sigma_\epsilon}\left(\frac{\partial^2 \epsilon}{\partial x^2}+\frac{\partial^2 \epsilon}{\partial y^2}\right)+C_{1\epsilon}\frac{\epsilon}{K}\mathcal{P}-C_{2\epsilon}\frac{\epsilon^2}{K}=0 \tag{5.36}$$

These equations are the basis of the heuristic method for generating the initial conditions for K and ϵ. (5.35)–(5.36) define a (coupled) boundary-value problem for K and ϵ, which is solved using a Newton-Raphson relaxation technique. The boundary conditions for (5.35)–(5.36) are the same as those for the full K-ϵ equations for $t>0$.

This heuristic method for generating the initial conditions for the turbulence variables gave good results in the grid turbulence, logarithmic-layer and steady turbulent cavity flow test problems (§ 5.11). In, general, it can be concluded that this method is acceptable when there is no experimental or theoretical solution for the specification of the initial conditions for K and ϵ.

5.9 Stability considerations.

In this section, the stability constraints for the 2D-HYDROTUR model are presented. The stability conditions reviewed in § 4.8 are extended for the irregular mesh, and new stability criteria for the K and ϵ equations are derived.

In 2D-HYDROTUR, stability constraints are imposed in two ways: off-line and on-line. The off-line analysis is performed by the user during the setup for a particular run; it is supported by a small program, "STAB.FOR", whose output is the appropriate cell dimensions and limiting Δt for the selected discretization (upwind, central, modified FTCS, or weighted combination of these schemes). The on-line analysis is performed by the program, in the form of an automatic Δt adjustment if the stability constraints are violated. This adjustment is done in a special subroutine (DELTAD), which also checks the number of pressure iterations per cycle, and reduces or increases Δt in order to keep that number between 10 and 25.

For accurate and stable computations, it is necessary to determine the appropriate cell dimensions Δx_i and Δy_j as functions of the expected velocities and effective viscosity, and the maximum admissible Δt for the selected spatial resolution. In this section, $\nu_e=\nu+\nu_t$ is the effective viscosity. When the turbulence model is not active $\nu_e=\nu$.

Determination of the spatial increments Δx_i and Δy_j

A quantitative estimate for the cell sizes Δx_i and Δy_j is provided by Hirt's criterion (§ 4.8.2)

$$\Delta x_{max}^2 \leq max\{2\nu_e / \left|\frac{\partial U}{\partial x}\right|\}_{i,j}$$

$$\Delta y_{max}^2 \leq max\{2\nu_e / \left|\frac{\partial V}{\partial y}\right|\}_{i,j}$$

Determination of Δt for the advection-diffusion components of the equations

Once the typical mesh sizes have been found using the above criterion, it remains to determine the maximum permissible Δt for the evolution equations—momentum, F-advection, and K-ϵ (if the turbulence model is "on"). The basic principles are :

a) The fluid cannot move through more than one cell in one time step.

b) Diffusion cannot be excessive, or deficient, in one time step.

c) The surface waves cannot move through more than one cell in one time step.

These principles are supplemented by the following heuristic criterion associated with the production and dissipation terms in the K-ϵ equations :

d) The relative variations of K and ϵ in one time step must be much smaller than unity.

Criteria a) and b) apply to the advection-diffusion components of the momentum and K-ϵ equations (neglecting the effects of pressure gradient in the former, and production and dissipation in the latter). For the numerical schemes used in the present model, these criteria are mathematically expressed by the inequalities presented in § 4.8.1 . These inequalities are now generalized for variable mesh sizes and viscosity.

If the convective term is discretized using central differences (FTCS scheme), the maximum permissible Δt must satisfy the following inequalities

$$\Delta t \leq min\{\frac{\Delta x}{|U|}, \frac{\Delta y}{|V|}\}$$

$$\Delta t \leq min\{\frac{\Delta x^2 \Delta y^2}{2\nu_e(\Delta x^2+\Delta y^2)}\}$$

$$\Delta t \leq min\{\frac{2\nu_e}{U^2+V^2}\}$$

These are the mathematical expression of a) and b) . For the modified FTCS scheme, the last two inequalities above are replaced by the following conditions

$$\Delta t \leq min\{\frac{[(\frac{\nu_e}{\Delta x^2}+\frac{\nu_e}{\Delta y^2})^2+(\frac{U}{\Delta x})^2+(\frac{V}{\Delta y})^2]^{1/2}-\nu_e(\frac{1}{\Delta x^2}+\frac{1}{\Delta y^2})}{(\frac{U}{\Delta x})^2+(\frac{V}{\Delta y})^2}\}$$

$$\Delta t \leq min\{\frac{2\nu_e}{|UV|}\}$$

If the convective term is discretized using the upwind scheme, the two conditions associated with the viscosity are replaced by the single inequality

$$\Delta t \leq min\{\frac{1}{2\nu_e(\frac{1}{\Delta x^2}+\frac{1}{\Delta y^2})+\frac{|U|}{\Delta x}+\frac{|V|}{\Delta y}}\}$$

which assures stability even when $\nu_e = 0$.

These stability conditions are only approximate, for they are based on a linear (Fourier-von Neumann) analysis, which does not include the simultaneous effects of the pressure gradient in the momentum equation, and the production/dissipation terms in the K-ϵ equations.

When a free-surface is included in the computation, the surface waves cannot be allowed to travel more than one cell in one time step. An estimate of the phase celerity of the surface waves is provided by the dispersion relation for linear waves. For shallow water waves, the Courant condition for the surface waves is

$$\Delta t \leq min\{\frac{\Delta x_i}{\sqrt{g h_{max}}}\}$$

where g is the acceleration due to gravity and h_{max} is the maximum fluid depth.

The above mentioned stability conditions apply to the convection-diffusion part of the momentum and K-ϵ equations. However, the K-ϵ equations pose additional stability problems, because of the stiff nature of the source-sink terms. The influence of these terms is now analyzed using criterion d) . This criterion was first introduced by Amsden & Harlow (1968) in a different turbulence model. The basic idea is to prevent a decay to negative values, which is physically impossible, if the dissipation is excessive.

Heuristic stability analysis of the source-sink terms in the K-ϵ equations

Heuristic stability conditions associated with the source-sink terms of the turbulence equations can be derived by neglecting the convection-diffusion part in the K and ϵ equations, and applying criterion d) (it is assumed that the stability conditions for the convection-diffusion part are independent of the source-sink terms, which of course is not true).

First, the special case $\mathcal{P}=0$ is considered. The simplified K-ϵ equations (which describe the decay of isotropic turbulence in a container) are

$$\frac{\partial K}{\partial t}=-\epsilon$$

$$\frac{\partial \epsilon}{\partial t}=-C_{2\epsilon}\frac{\epsilon^2}{K}$$

In these expressions, $(\epsilon/K)^{-1}$ is a characteristic decay time, here denoted by $\mathcal{T}$. Physically, $\mathcal{T}$ is a characteristic decay (or "turn-over") time for the large scale eddies. In terms of $\mathcal{T}$, the simplified K-ϵ equations read

$$\frac{\partial K}{\partial t}=-\frac{K}{\mathcal{T}}$$

$$\frac{\partial \epsilon}{\partial t}=-C_{2\epsilon}\frac{\epsilon}{\mathcal{T}}$$

which show the similarity of the dissipation terms in the K and ϵ equations. For the relative variation of K and ϵ to be small in one time step, it is necessary to verify the following two conditions

$$\left|\frac{\Delta K}{K}\right|=\frac{1}{\mathcal{T}}\Delta t\ll 1 \quad ; \quad \left|\frac{\Delta\epsilon}{\epsilon}\right|=C_{2\epsilon}\frac{1}{\mathcal{T}}\ll 1$$

or

$$\Delta t\ll min\{\frac{K}{\epsilon}\} \tag{5.37}$$

$$\Delta t\ll min\{\frac{1}{C_{2\epsilon}}\frac{K}{\epsilon}\} \tag{5.38}$$

These conditions are very similar, but the condition resulting from the simplified ϵ equation is slightly more restrictive because $C_{2\epsilon}=1.92$. Therefore, the ϵ equation is "stiffer" than the K equation, *i.e.* it is more difficult to stabilize (this fact is well known in the literature of turbulence modeling). Conditions (5.37)-(5.38) show that even though the dissipation term of the K equation is not singular when $K\rightarrow 0$, it can originate the same type of instability associated with the singular terms of the ϵ equation. If these conditions are not verified, a decay of K or ϵ to negative values can occur. This is physically impossible, and produces a negative eddy-viscosity which rapidly destroys the calculation. There is no simple criterion for the maximum $\Delta t/\mathcal{T}$. In the present work, values ranging from 10^{-3} to 10^{-2} were used, depending on the other stability conditions and the characteristics of the flow to be simulated.

From the above heuristic stability analysis, the following was concluded :

- The similarity between the K and ϵ equations was highlighted.

- The numerical problem of stability is related with the physical problem of breakdown of the standard K-ϵ model when $K/\epsilon \rightarrow 0$. In this case, turbulence cannot be maintained, and there is no Δt which prevents a decay to negative values. K/ϵ is the relevant parameter governing the stability of the simplified turbulence equations (5.37)-(5.38).

If the production term is included in the analysis, it is found that the limiting Δt should satisfy the following inequalities

$$\Delta t \ll \mathcal{T}\frac{\mathcal{T}^*}{|\mathcal{T}^* - \mathcal{T}|} = \mathcal{T} f(\mathcal{T}^*, \mathcal{T})$$

$$\Delta t \ll \mathcal{T}\frac{\mathcal{T}^*}{|C_{2\epsilon}\mathcal{T}^* - C_{1\epsilon}\mathcal{T}|} = \mathcal{T} g(\mathcal{T}^*, \mathcal{T})$$

where $\mathcal{T}^* = K/\mathcal{P}$ is a "characteristic feeding time". If $\mathcal{T}^* \rightarrow \infty$ ($\mathcal{P} \rightarrow 0$, K fixed), then $f(\mathcal{T}^*, \mathcal{T}) \rightarrow 1$ and $g(\mathcal{T}^*, \mathcal{T}) \rightarrow 1/C_{2\epsilon}$, and (5.37)-(5.38) are recovered. Physically, $\mathcal{T}^* \rightarrow \infty$ implies that turbulence cannot be maintained because it takes a very long time to extract energy from the mean flow. If $\mathcal{T}^* \rightarrow 0$, *i.e.* if $\mathcal{P}$ (and the velocity gradient of the mean flow) is large with $\mathcal{T}$ finite, the heuristic stability conditions are $\Delta t \ll \mathcal{T}^*$ and $\Delta t \ll 1/C_{1\epsilon}\mathcal{T}^*$. In this case the small Δt is required to prevent a rapid growth of the solution, rather than a fast decay to negative values. When the turbulence is near to equilibrium, *i.e.* $\mathcal{P} \simeq \epsilon$, $\mathcal{T}^* = \mathcal{T}$ and the source-sink terms do not originate instability problems (the associated heuristic stability condition is simply $\Delta t < \infty$).

5.10 Programming considerations.

2D-HYDROTUR was constructed in a highly modular form using the rules of structured programming (Dijkstra, 1976; Smith, 1982; Welland, 1986). First, the structure of the data was defined. Shared data areas (named COMMON blocks) were created according to the functional role of the variables: mesh-related variables, global variables (constants, index bounds, flags and counters), field variables for the NS solver, and field variables for the turbulence model.

The subprogram modules of 2D-HYDROTUR are highly cohesive, *i.e.* each module is an integral part, essential to the performance of a single function, and does not interfere with the other modules. Coupling between modules is done via global variables and shared data areas. Systematic testing was performed during all stages of development, using three basic principles (Welland, 1986): incremental testing, black-box testing and glass-box testing. After constructing the main driver program, all subprograms were first replaced by dummy modules (stubs). After the module performed as expected, the corresponding dummy module was deleted.

Optimization is an important factor in the present model, because a total of six partial differential equations are integrated simultaneously in a time varying domain. The computer program has been optimized for both scalar and vector processing. Therefore, it makes full use of the capacity of the machine in which it is installed. Scalar optimization consisted of minimizing the number of operations in the most time-consuming parts of the program. Optimization for vector and parallel processing is much more difficult, and requires access to a vector/parallel computer. The vector and parallel optimization was performed in the following steps :

- Profile analysis of the total execution time consumed by each subroutine, and by each loop within the various subroutines.

- Compiler and hand analysis of the loops that were automatically vectorized or rejected (due to data dependencies such as feedback to array elements, potential multiple store, non-vectorizable conditionals, non-vectorizable subroutine calls, complex branching, etc).

- Total or partial recodification of the loops in the most time consuming routines.

The process of code optimization for vector processing is now illustrated for the case of the pressure-velocity iteration step, which in the scalar version consumes about 60% of the total execution time. The structure chart for this subprogram is shown in figure 5.9 .

In the main loop, the current cell is identified as surface or full by testing its F value. For the surface cells, there exists a "steering flag" NF(I,J) , whose value is 1, 2 , 3 or 4 if the interpolation (full) cell is on the left, right, bottom or top, respectively. To find the interpolation cell, NF(I,J) is tested four times.

The main loop in figure 5.9 does not vectorize because of data dependency in both I and J indices for U, V, and φ. Also, there are too much IF statements inside the loop (mainly those for testing the free-surface configuration in order to decide the correct expression for $\Delta\varphi$), and this reduces vector performance considerably.

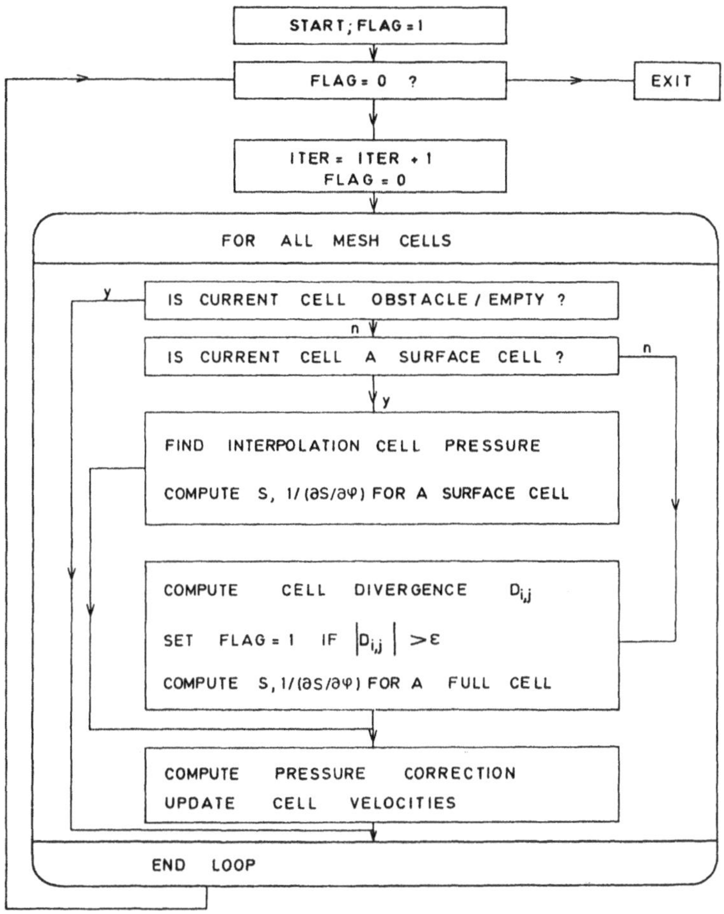

Figure 5.9 : *Structure chart for the pressure-velocity iteration step.*

Data dependency was eliminated by splitting the main loop into four smaller loops with an increment of 2, *i.e.* :

```
DO I=2, IMAX-1, 2
DO J=2, JMAX-1, 2
:
ENDDO
DO I=3, IMAX-1, 2
DO J=2, JMAX-1, 2
:
ENDDO
```

```
DO  I=2, IMAX-1, 2
DO  J=3, JMAX-1, 2
  :
ENDDO
DO  I=3, IMAX-1, 2
DO  J=3, JMAX-1, 2
  :
ENDDO
```

each containing the same instructions. This eliminates recurrence of the form U(I,J)=U(I+1,J) or U(I,J)=U(I-1,J) . This technique is normally used to vectorize Laplace and Poisson solvers, and was adapted in the present work to the pressure-velocity iteration. The problem of excessive conditionals inside the loops was solved by converting them in arithmetic operations, which are much faster for vector processing. Instead of testing for NF(I,J) to set the interpolation pressure φ_f , φ_f is computed without IF statements using the expression :

$$\varphi_f = -\frac{(x-2)(x-3)(x-4)}{6}\varphi_{i-1,j} + \frac{(x-1)(x-3)(x-4)}{2}\varphi_{i+1,j} + \\ -\frac{(x-1)(x-2)(x-4)}{2}\varphi_{i,j-1} + \frac{(x-1)(x-2)(x-3)}{6}\varphi_{i,j+1} \tag{5.39}$$

with $x = \mathrm{NF(I,J)}$. The four products in the numerators have two common factors, and this was taken into account to reduce the number of multiplications involved in the selection of the interpolation pressure. Using these optimization techniques, the time consumed by the pressure calculation was reduced by a factor of 4.5 . The total execution time was reduced by a factor of 2.5 , which is a good speed factor for a 3-processor ALLIANT FX-40 computer where the vectorized version of the code was prepared. The explicit momentum calculation and the whole turbulence model were also vectorized, by choosing adequate loop structures and inserting compiler directives.

5.11 Selected test problems.

The numerical model here developed was tested using the following types of flow: laminar cavity flow at high *Re*, laminar dam-break (collapse of a fluid column) problem, turbulence decay behind a grid, logarithmic turbulent boundary layer, and turbulent cavity flow. For all test problems, either analytical or numerical results were available for comparison.

Laminar cavity flow is an important type of confined flow, with fascinating properties. It was included to test the ability of the model for representing the instabilities and bifurcations that arise in the solutions of the NS equations (which ultimately end in turbulence, if the Reynolds number is high enough). The dam-break problem is a well studied type of free-surface flow, which provides a stringent test on the capacity of MAC-type models for treating highly contorted free-surfaces. The decay of turbulence behind a grid is a comparatively simple turbulent flow which cannot be treated by means of algebraic closures. The simplified K-ϵ equations provide analytical expressions for the spatial distribution of the turbulence variables, with which the numerical results were compared. The logarithmic boundary layer plays a central role in the theory of wall flows, and in the present model. A simulation of the classical log-layer profiles was included to test performance of the model when turbulence is near to the equilibrium condition $\mathcal{P} = \epsilon$, and as a way of validating the present treatment of the wall boundary conditions. Finally, in the turbulent cavity flow problem, all turbulent transport processes were active, and the governing equations could not be simplified to provide analytical solutions. However a numerical investigation by Chen & Chang (1987) was available for comparison. In the confined flow problems, the initial and boundary conditions were chosen so that $F = 1$ everywhere for $t \geq 0$. The numerical model represented accurately the features of the solutions in all test problems. All simulations reported in this section were done using an IBM-PC compatible computer. This shows that many interesting and varied fluid flow problems can be investigated using 2D-HYDROTUR, without the need for large computers.

5.11.1 Laminar cavity flow.

The flow inside a cavity driven by the uniform translation of the upper boundary plays an important role in the class of confined flows with separation. It has been object of intense investigation (see Gustafson & Halasi, 1986, 1987 and Goodrich & Soh, 1989). The numerical studies indicate that a large primary vortex is formed near the geometric center of the cavity, together with secondary and tertiary vortices near the corners. The pattern of the secondary and tertiary vortices is a function of the Reynolds number. The complexity of the flow topography increases with Re, and is probably due to bifurcations in the solutions of the NS equations (Gustafson, 1987).

Two flows in a unit square cavity with Re equal to 1000 and 2000 were simulated, to demonstrate the ability of the numerical scheme for representing the formation of the secondary flow patterns at high Reynolds numbers. The flow topography was enhanced by dividing the velocity by its magnitude and providing perspective plots of kinetic energy and vorticity. The boundary conditions were chosen as follows. At the top boundary, $U = 1\ ms^{-1}$ and $V = 0\ ms^{-1}$. At the remaining boundaries, no-slip conditions were prescribed. The solution was started from rest and the NS equations were marched in time until a steady solution was reached. The body force was neglected in both cases. In the first test ($Re = 1000$) a mesh of 30×30 regular cells with $\Delta x, \Delta y = 1/30$ was used. In the second test ($Re = 2000$) a mesh of 40×40 regular cells with $\Delta x, \Delta y = 1/40$ was used.

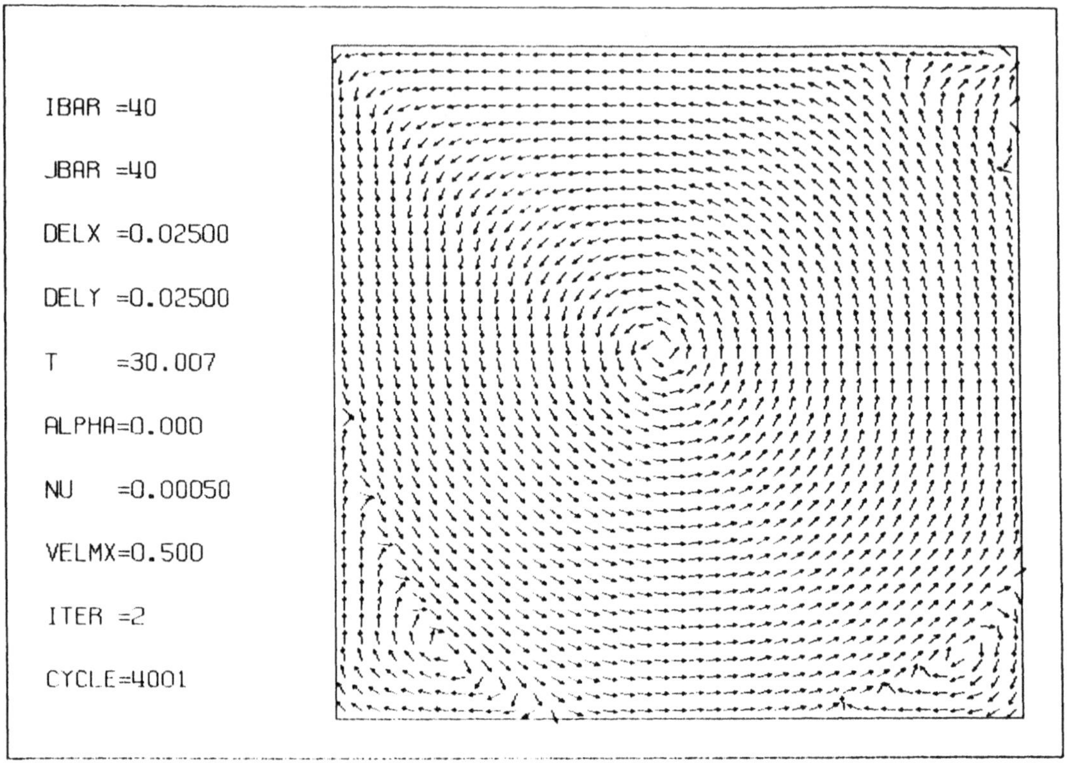

Figure 5.10 : *Normalized velocity field in a driven cavity flow.* $Re = 2000$, $t = 40$ s .

Figures 5.10–5.12 show the computed velocity, kinetic energy and vorticity fields for $Re = 2000$. The position, strength and spatial extension of the large primary vortex and of the secondary corner vortices were in good agreement with published results (Gustafson & Halasi, 1986, 1987; Goodrich & Soh, 1989). The onset of separation close to the top right corner was observed, where a tertiary vortex was formed (this tertiary vortex was not seen in the $Re = 1000$ simulation). This is also in agreement with the previous studies. The kinetic energy plot showed that the dynamics of the secondary corner vortices can only be perceived if the velocity field is normalized, for the high kinetic energy is concentrated in the primary vortex. The vorticity plot shows other interesting features of the problem. First, the vorticity is strongly concentrated close to the boundaries, as was expected. Indeed, the whole flow pattern is generated by the diffusion of vorticity from the top boundary. Second, the small valleys of the vorticity surface near the bottom corners roughly coincide with the separation streamlines.

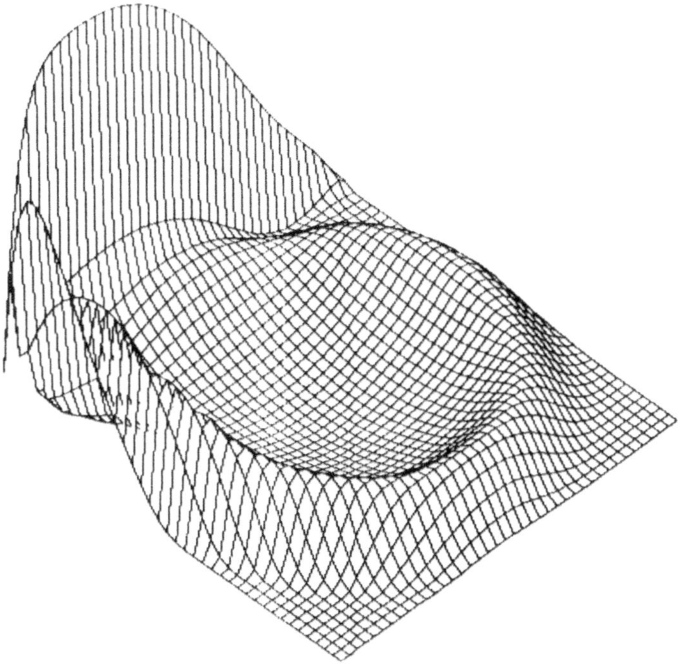

Figure 5.11 : *Kinetic energy surface in a driven cavity flow.* $Re = 2000$, $t = 40$ s .

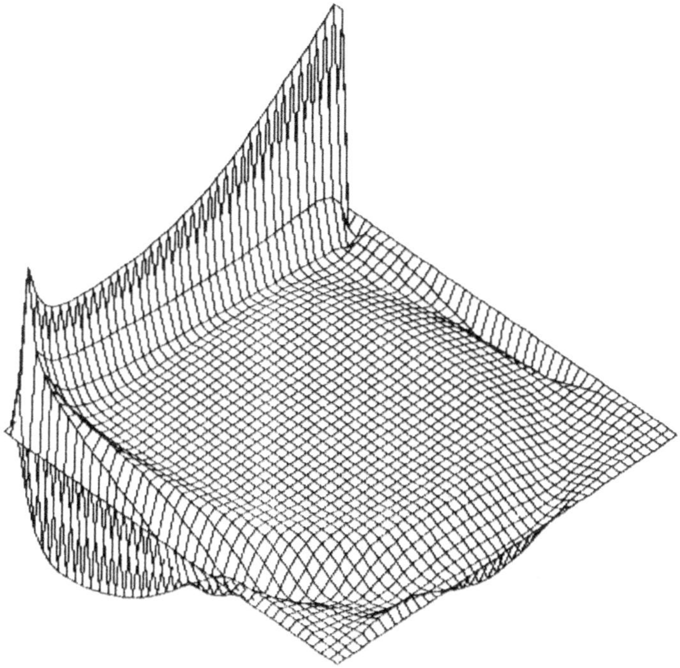

Figure 5.12 : *Vorticity surface in a driven cavity flow.* $Re = 2000$, $t = 40$ s .

5.11.2 Grid turbulence.

When a uniform stream of fluid in laminar regime is made to pass through a grid, turbulence is generated at the grid position. This type of turbulence is nearly isotropic and decays downstream of the grid location in the absence of production (Batchelor, 1953). The resulting turbulent flow cannot be described using algebraic closures of the mixing length type, and provides a good test problem for the K-ϵ model. Let U_g be the uniform stream velocity. If $U_g d/dx \gg d/dx(\nu_t d/dx)$ the distribution of $K=K(x)$ and $\epsilon=\epsilon(x)$ is given by the following simplified K-ϵ equations

$$U_g\frac{dK}{dx} = -\epsilon \tag{5.40}$$

$$U_g\frac{d\epsilon}{dx} = -C_{2\epsilon}\frac{\epsilon^2}{K} \tag{5.41}$$

The general solution of (5.40)–(5.41) is

$$K = K_g(1+A_1\frac{\epsilon_g}{K_gU_g}x)^{-n}$$

$$\epsilon = \epsilon_g(1+A_1\frac{\epsilon_g}{K_gU_g}x)^{-(n+1)}$$

where the subscript $_g$ denotes values at the grid. In the present test problem, $U_g=1\ ms^{-1}$ without loss of generality. Substituting into (5.40)–(5.41) it is found that

$$\frac{K}{K_g} = (1+0.91\frac{\epsilon_g}{K_g}x)^{-1.1} \tag{5.42}$$

$$\frac{\epsilon}{\epsilon_g} = (1+0.91\frac{\epsilon_g}{K_g}x)^{-2.1} \tag{5.43}$$

are the approximate decay laws for K and ϵ. In the numerical calculation a regular mesh of 30×15 cells ($\Delta x,\Delta y=0.1$) was used. At the left boundary an inflow velocity of 1 ms^{-1} was specified. The inflow condition for the turbulence variables was $K_g=5\times10^{-2}\ m^2s^{-2}$ and $\epsilon_g=4.5\times10^{-2}\ m^2s^{-3}$. The remaining boundaries were treated as outflow boundaries for all variables. The initial condition for the mean velocity was $U=1\ ms^{-1}$ everywhere, and the initial conditions for the turbulence variables were generated using the heuristic method of § 5.8 . The full unsteady equations were marched until a steady spatial distribution was reached (this occurred for $t\simeq5$ s).

Figure 5.13 shows the computed *vs* analytical decay of K and ϵ behind the grid as a function of position. The velocity field remained exactly uniform (to the machine round-off error) during the whole calculation. The agreement between the computed and the analytical values of the turbulence variables was extremely good. Furthermore, all variables remained uniform in y as they should. This calculation took 2 minutes of computer time in an IBM-PC compatible computer with a performance of 24 KFLOPS.

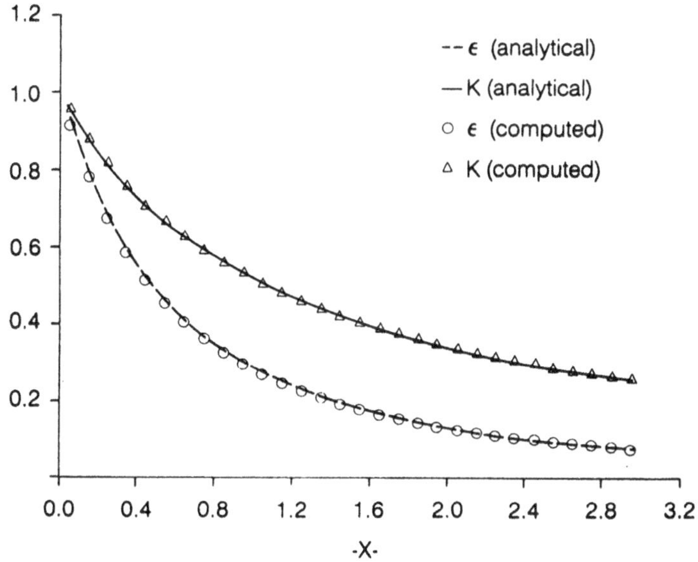

Figure 5.13 : *Analytical vs computed decay of turbulence intensity K, and its rate of dissipation ϵ behind a grid.*

5.11.3 Logarithmic boundary layer.

In this section, the theoretical profiles of U, K, ϵ and ν_t for a logarithmic turbulent boundary layer over a smooth wall are compared with numerical results obtained with 2D-HYDROTUR. This study was particularly important for the validation of the boundary condition treatment.

The height and length of the solution domain were chosen 1 m and 10 m, respectively. The domain was discretized using a regular mesh of 10×10 cells. The molecular viscosity and the friction velocity were chosen $\nu = 10^{-6}\ m^2s^{-1}$ and $u_* = 10^{-2}\ ms^{-1}$, respectively. The logarithmic velocity profile was then

$$U = 2.44\times10^{-2} ln(9.0\times10^{3}y) \quad (5.44)$$

with $E = 9.0$ and $\kappa = 0.41$. At the top boundary, U, K, and ϵ were prescribed using (5.44),

$K = u_*^2/(C_d)^{1/2}$ and $\epsilon = u_*^3/0.41$. The left and right boundaries were specified open. The bottom boundary was coincident with the wall. Consequently, default settings consistent with the law-of-the-wall were chosen for all variables. The velocity field was initialized uniform, with $U = U(y=1)$ .The initial conditions for K and ϵ were generated using the heuristic method of § 5.8 . The choice of the boundary conditions was intended to test the performance of the wall and open boundary treatments. At the same time, the open boundary conditions should allow correct entrance and exit of the fluid.

Figures 5.14–5.17 show the computed *vs* analytical profiles for U, K, ϵ and ν_t/ν (the local turbulence Reynolds number). There was good agreement between the computed and the analytical profiles. However there were some discrepancies. To investigate the origin of these discrepancies, the value of the various terms of the discretized equations was monitored during the execution. It was concluded that the term $\mathcal{P}-\epsilon$ was about 5% of $\mathcal{P}$ or ϵ. Therefore, the model was not representing exact equilibrium. This happens because when $(\mathcal{P}-\epsilon)_k = 0$, $(\mathcal{P}-\epsilon)_\epsilon \neq 0$ and vice versa. The difference between production and dissipation was balanced by cross-stream diffusion, which is not taken into account in the theoretical model. The discrepancies in the profiles of ν_t/ν and ϵ close to the wall were due to the fact that the computed tangential velocity (consequently the friction velocity) was higher than the theoretical value. It would have been possible to reduce this difference by increasing the duration of the simulation. However, because the flow was near to equilibrium, any changes would have been very slow. This simulation provided a fair validation of the boundary condition treatment. The open boundaries allowed the correct entry and exit of fluid, whereas the wall boundary conditions provided a correct transmission of the shear stress to the layer of fluid.

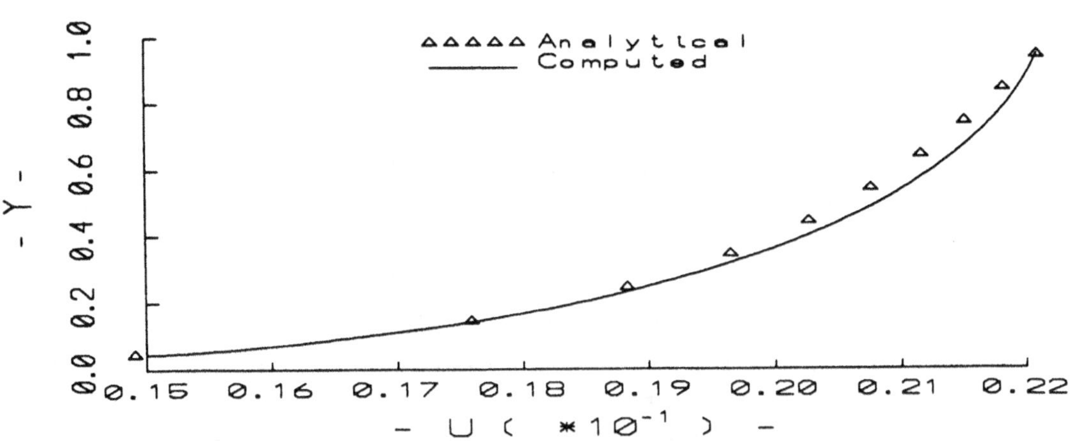

Figure 5.14 : *Computed vs analytical velocity profile for a logarithmic boundary layer. Solid line: computed solution. Triangles: analytical solution.*

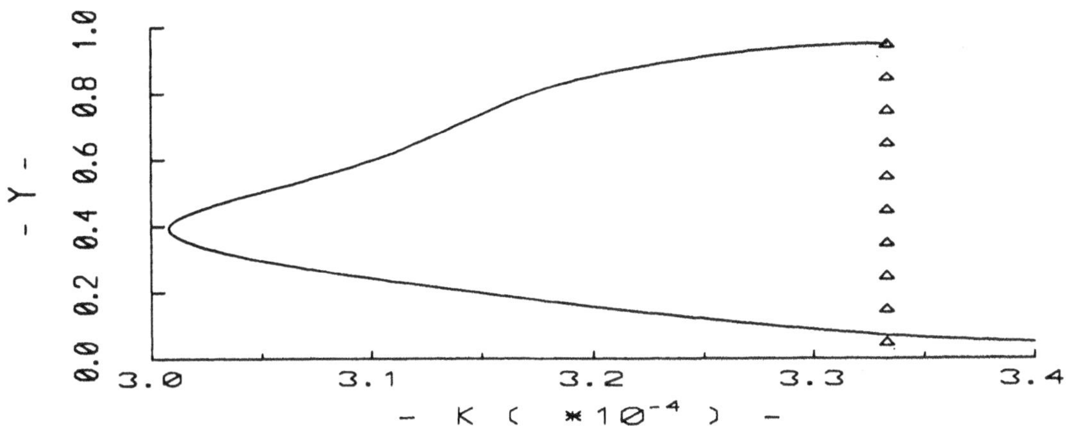

Figure 5.15 : *Computed vs analytical turbulent kinetic energy (K) profile for a logarithmic boundary layer. Solid line: computed solution. Triangles: analytical solution.*

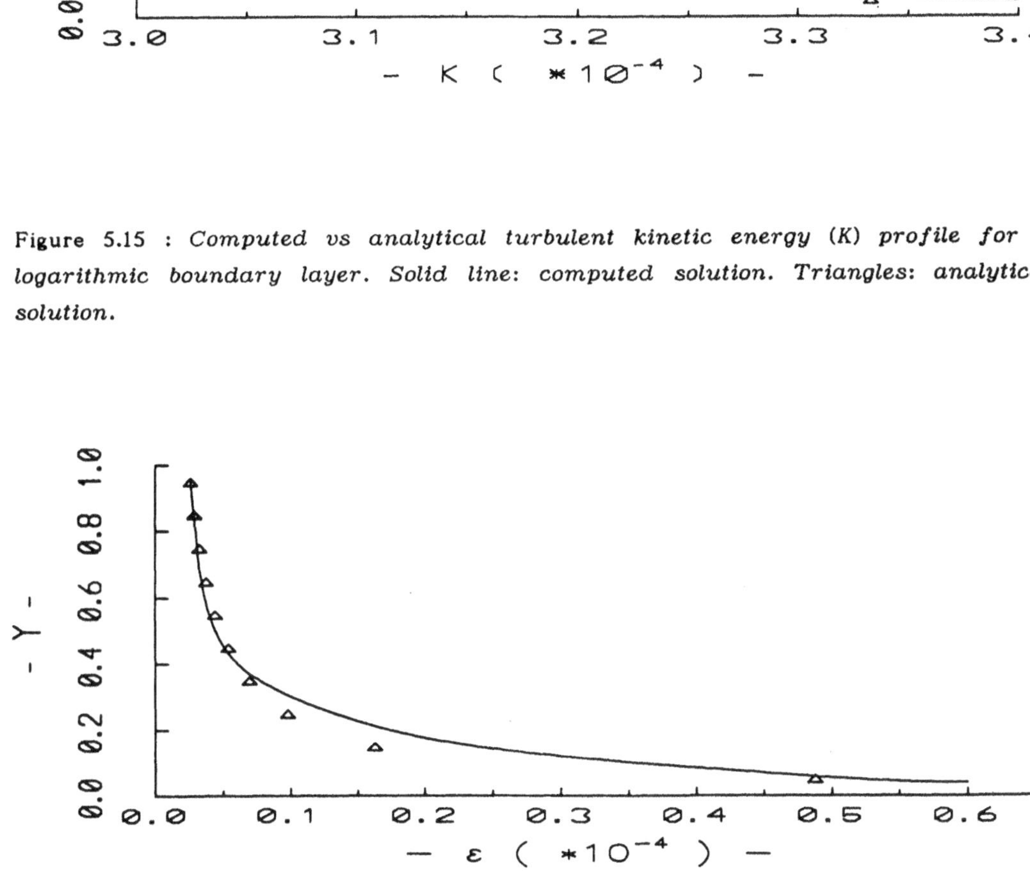

Figure 5.16 : *Computed vs analytical rate of dissipation (ϵ) profile for a logarithmic boundary layer. Solid line: computed solution. Triangles: analytical solution.*

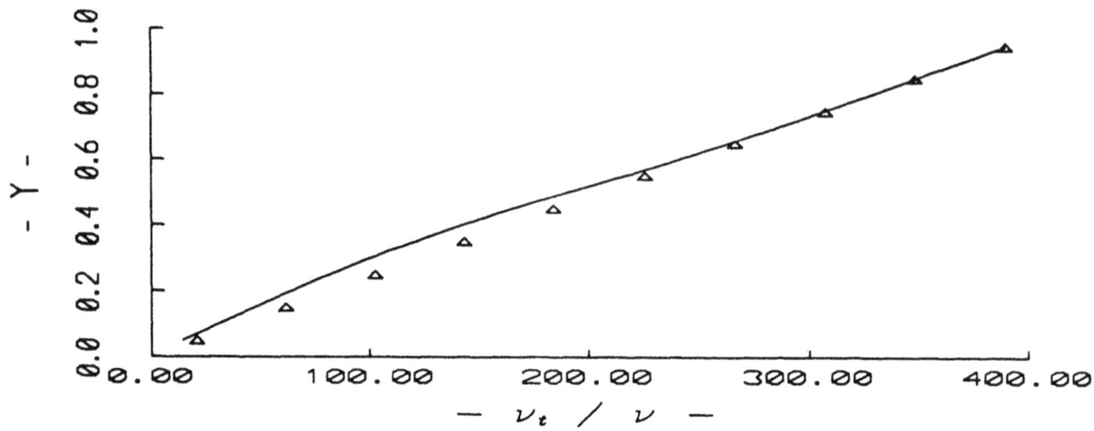

Figure 5.17 : *Computed vs analytical ν_t/ν (local turbulence Reynolds number) profile for a logarithmic boundary layer. Solid line: computed solution. Triangles: analytical solution.*

5.11.4 Turbulent cavity flow.

In this section, the problem of § 5.11.1 is analyzed from a different viewpoint. As the Reynolds number increases, the flow becomes turbulent. The no-slip condition cannot be used, and is replaced by the partial slip/law-of-the-wall prescription for the mean velocity. There are strong velocity gradients within the flow, and production and dissipation of turbulence are not in equilibrium. Chen & Chang (1987) reported a calculation of turbulent cavity flow at $Re = 10^5$ using the K-ϵ model, with which the present results were compared. They found that the mean velocity field is markedly different from the velocity field in a laminar cavity flow.

To compare the results of the present model with those of Chen & Chang (1987), the Reynolds number (based on the length and translation velocity of the top lid) was set to 10^5. The velocity boundary conditions were the following. At the top boundary, a tangential velocity of 1 ms^{-1} was specified, and the normal velocity was set to zero. At the remaining boundaries, partial-slip conditions consistent with the law-of-the-wall were specified. The boundary conditions for K and ϵ at the left, right and bottom boundaries were generated using the law-of-the-wall. For the top boundary, the law-of-the-wall was also used, but the friction velocity was computed according to the expression

$$|U_t - U_{wall}|/u_* = (1/\kappa)\,ln(Ey^+)$$

where U_t is the tangential velocity at the cell near to the boundary and $U_{wall} = 1\ ms^{-1}$ is the velocity of the top lid. A regular mesh of 20×20 cells (nearly 1/2 of the resolution of Chen & Chang, 1987) was used. The initial conditions were generated as follows. For the velocity field, a laminar solution with a viscosity of $10^{-3}\ m^2s^{-1}$ was used. This initial solution was obtained by running the model without the K-ϵ equations for a few cycles using free-slip conditions at the left, right and bottom boundaries. The turbulence field was initialized using the heuristic method described in § 5.8 .

Figure 5.18 represents the computed velocity field and the contour lines of ν_t/ν for $t = 60\ s$, when the flow reached a steady condition. Figures 5.19 and 5.20 illustrate the computed contour lines for K and ϵ at the same instant.

Instead of secondary and tertiary vortices, a large single vortex was observed, with center close to the geometric center of the cavity. There was large turbulence production near the top right corner, where local maxima of ν_t , K and ϵ appeared. The modulus and spatial distribution of the contour lines were in good agreement with the results of Chen & Chang (1987) (which do not present contour plots of ϵ for comparison). This confirms the global performance of the present K-ϵ model in this type of recirculating flow.

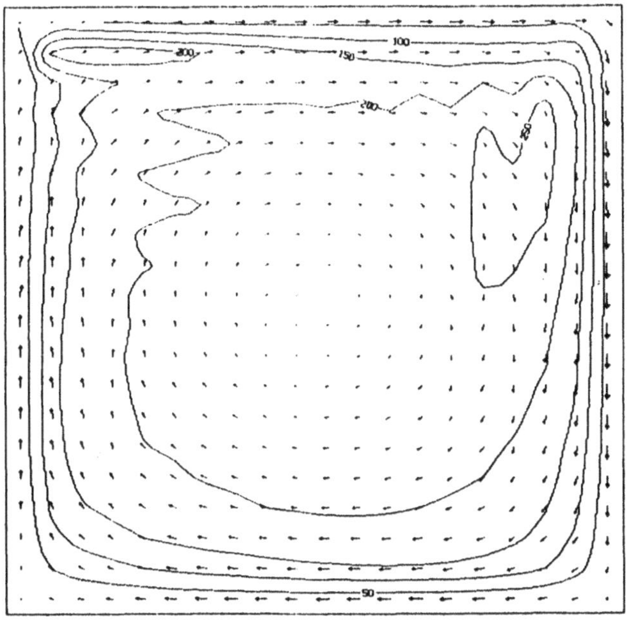

Figure 5.18 : *Computed velocity field and contour lines of* ν_t/ν , $t = 60\ s$.

Even though the initial distribution of the turbulence variables was qualitatively different from the final one, the heuristic method produced a good initial condition for the turbulence field. The local maxima of K and ϵ were already located close to their final positions. Other simpler initial conditions required a larger time interval to reach the steady solution, and in some cases generated important instabilities.

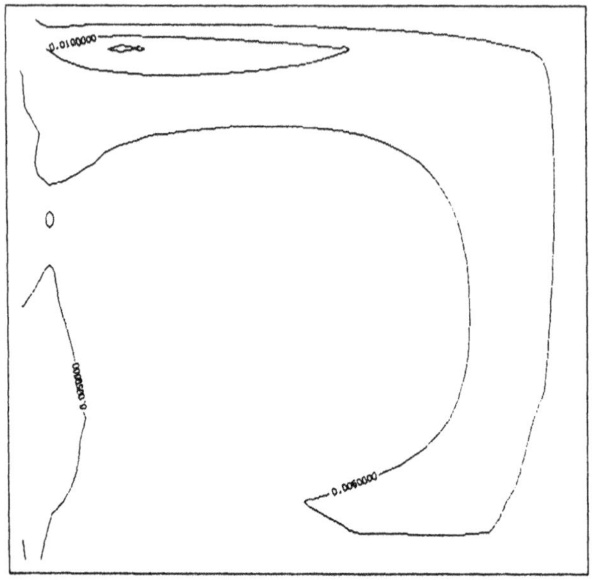

Figure 5.19 : *Computed contour lines of turbulent kinetic energy K, t = 60 s .*

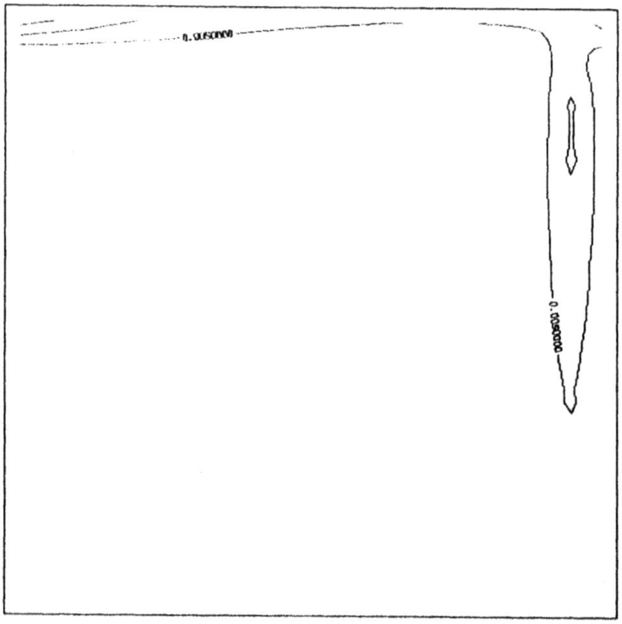

Figure 5.20 : *Computed contour lines of turbulence rate of dissipation* ϵ, *t* = 60 *s* .

This simulation was performed using an IBM-PC compatible computer. The performance was 24 KFLOPS, and the total execution time was 5.5 hours. The K-ϵ equations increased the computational cost, but the higher effective viscosity allowed a larger time step than was possible in the high *Re* cavity flow simulations of § 5.11.1 .

5.11.5 Dam-break problem.

The collapse of a column of fluid under the action of gravity is a classical test problem for MAC-type algorithms (Welch et al. , 1966; Amsden & Harlow , 1970; Nichols et al. , 1980). This problem was simulated with the numerical model here developed, and the results were compared with those of Amsden & Harlow (1970). To demonstrate the auxiliary features of the program and the accuracy of the VOF method, a set of marker particles was included in the calculation.

The data for this calculation were taken from Amsden & Harlow (1970). Figure 5.21 illustrates the solution domain and the initial fluid configuration. The solution domain was a rectangular cavity 15 m long and 7.5 m high, which was discretized using a rectangular mesh of 30×15 regular cells. The fluid column was 7 m high and 3.5 m long. The boundary conditions were chosen free-slip for all boundaries, and the molecular viscosity was chosen $\nu = 0.2\ m^2 s^{-1}$ (following Amsden & Harlow, 1970).

Figures 5.22–5.24 represent the instantaneous velocity field and fluid configuration for the times $t = 3$ s, $t = 6$ s and $t = 15$ s , respectively. Figures 5.25–5.27 represent the marker particle configuration for the times $t = 3$ s , $t = 6$ s , and $t = 15$ s , respectively.

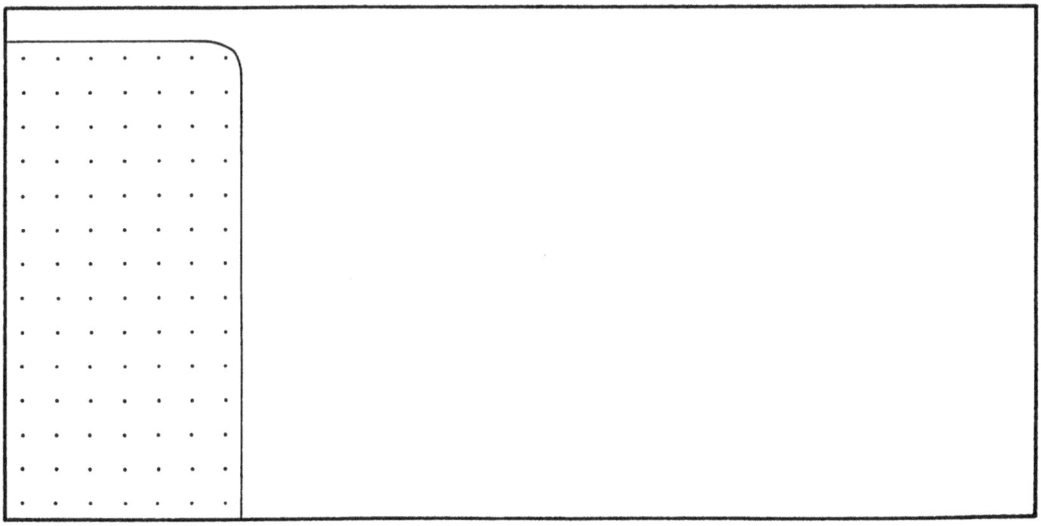

Figure 5.21 : *Initial fluid configuration for the dam-break problem.*

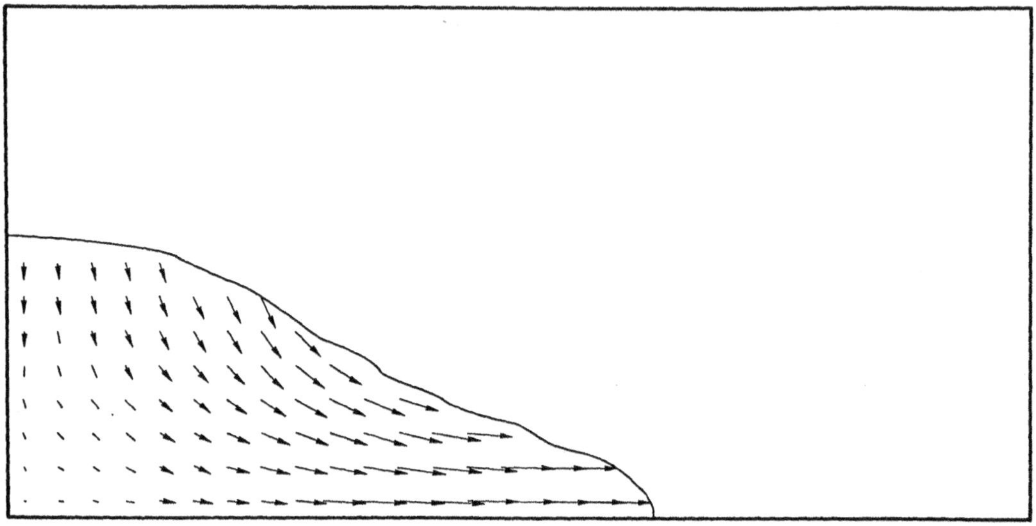

Figure 5.22 : *Velocity field and free-surface configuration for the dam-break problem*, $t = 3.0$ s .

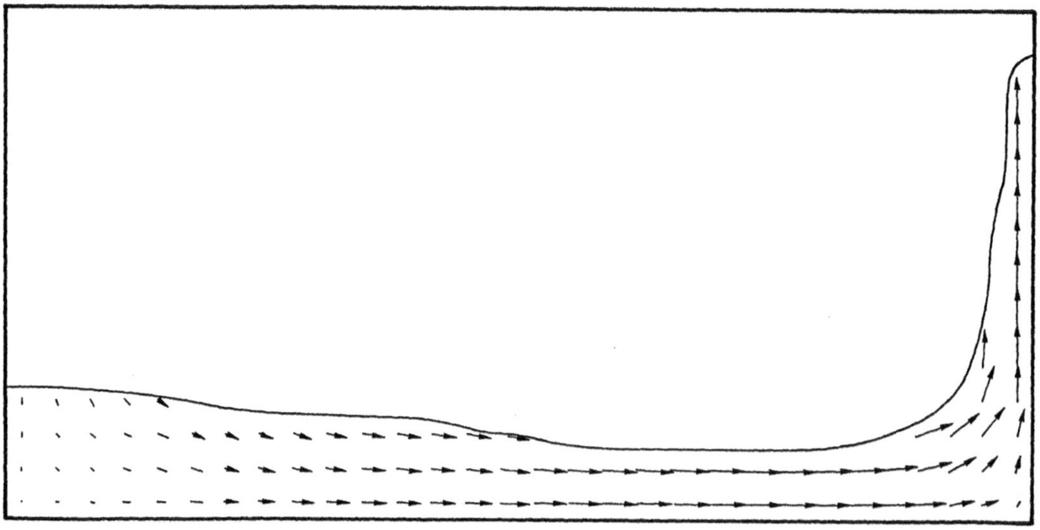

Figure 5.23 : *Velocity field and free-surface configuration for the dam-break problem*, $t = 6.0$ s .

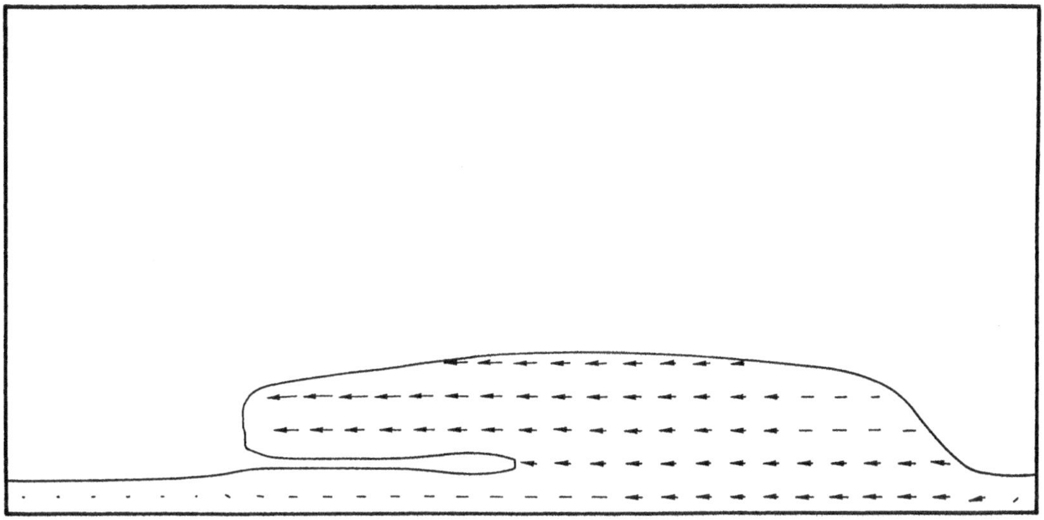

Figure 5.24 : *Velocity field and free-surface configuration for the dam-break problem*, $t = 15.0$ s .

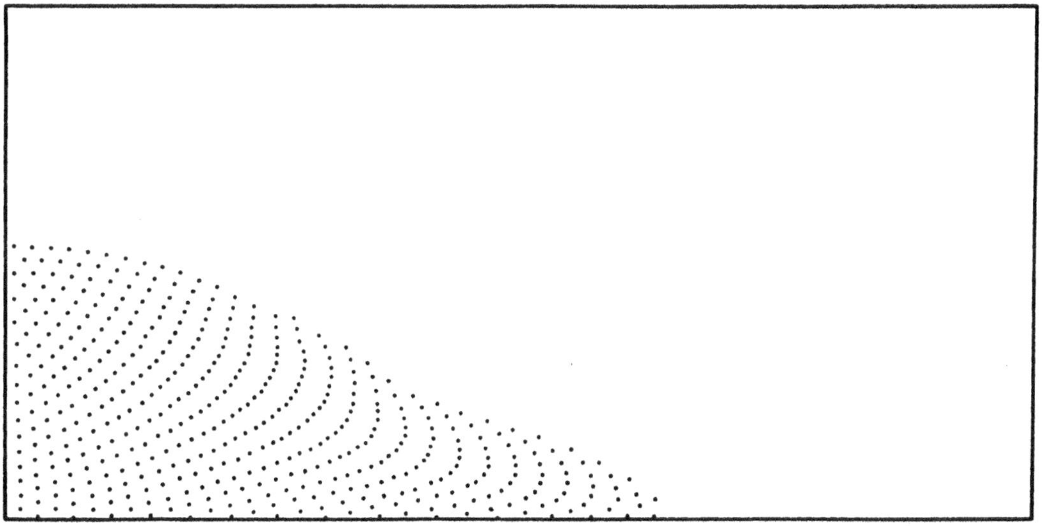

Figure 5.25 : *Marker particle configuration for the dam-break problem*, $t = 3.0$ s .

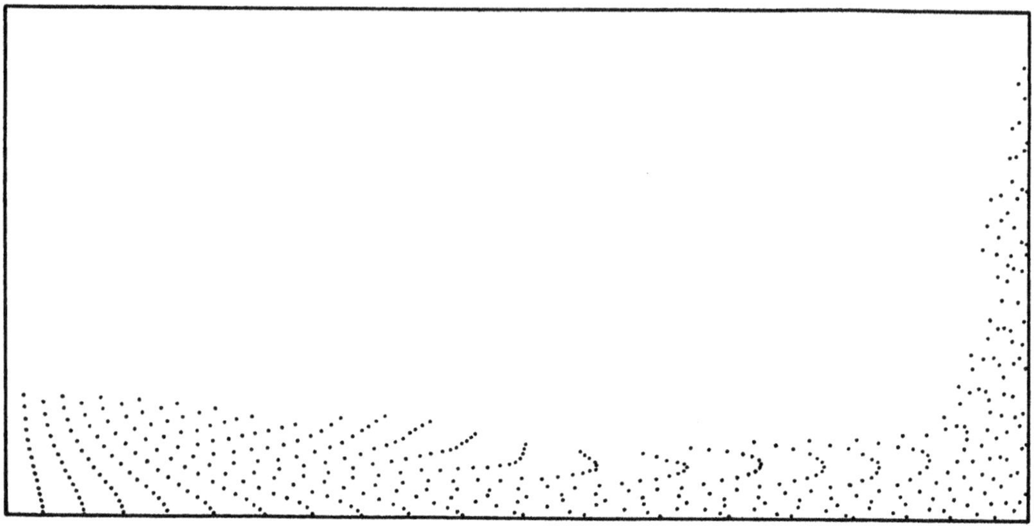

Figure 5.26 : *Marker particle configuration for the dam-break problem,* $t = 6.0$ s .

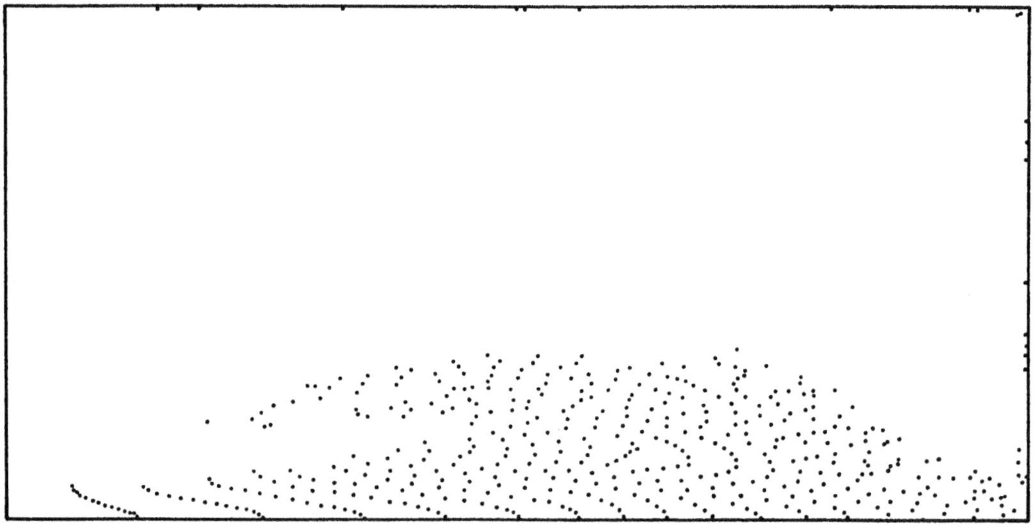

Figure 5.27 : *Marker particle configuration for the dam-break problem,* $t = 15.0$ s .

Chapter 6

NUMERICAL SIMULATION OF SHALLOW WATER WAVES

6.1 Introduction.

The purpose of this chapter is to illustrate the capabilities of the present numerical technique for calculating wave transformation problems. The emphasis will be on the unified treatment of the flow before, during and after breaking. Several typical wave transformation problems were selected, and the results of the computations will be presented by increasing order of complexity. From the viewpoint of surf-zone hydrodynamics, the complexity of the physical phenomena increases from pre-breaking, to transitional (overturning), and finally to broken (turbulent) motion. The numerical simulations reported below are ordered according to this sequence.

First, pre-breaking phenomena are considered, by studying the propagation and collision of solitary waves (§§ 6.2–6.3). The numerical solutions showed that the present model describes accurately the evolution of highly nonlinear waves when nonlinear steepening and dispersivity are in balance. It was found that the order of the theory used for generating initial and boundary conditions has a greater influence in the quality of the simulations than the truncation errors. Second, transition from dispersive to dissipative regime was investigated using the moving hydraulic jump as the basic model (§ 6.4). This study was divided in three stages—undular, transitional and turbulent jumps—corresponding to increasing Froude numbers and depth ratios. It was shown that the three types of jump have distinctive characteristics, which in turn are similar to those of pre-breaking, overturning, and broken waves. The properties of undular jumps were predicted with great accuracy. The computed critical depth ratio was found to agree with the experimentally observed value, and the nature of the transition was explained in terms of the structure of the velocity field. The main features of the turbulence field in turbulent jumps were accurately reproduced, except very near the jump's front and close to the bottom.

Finally, the typical sequence of wave transformation phenomena across a surf-zone was simulated using both solitary and periodic waves (§ 6.5–6.6). A qualitative and quantitative discussion of the numerical solutions was made using the experimental results of Nadaoka (1986) and Mizuguchi (1986).

6.2 Propagation of a solitary wave over a horizontal bottom.

The propagation of a solitary wave over a horizontal bottom is a classic problem for testing nonlinear models. The accuracy of the model can be evaluated by comparing the computed celerity and amplitude of the solitary wave with their theoretical values, and by observing the deformation of the original wave during the propagation. In the numerical simulations reported below, wave theories of first, second and third order were used. The effects of the order of approximation, truncation errors and F-convection algorithm will be discussed.

The physical parameters of the problem were chosen as follows. First, time and length units were chosen so that the acceleration of gravity, g, and the still water depth, D, were both equal to 1. A solitary wave with height $H=0.4$ and initial crest position $x_{crest}=15.0$ was placed in a domain 40 units long and 2 units high. The bottom boundary was chosen free-slip and the remaining boundaries were specified open. On the free-surface inviscid boundary conditions were used.

In the x-direction, the domain was discretized using a uniform mesh of 160 cells with $\Delta x=0.25$. In the y direction, a variable mesh of 18 cells with minimum spacing $\Delta y_j=0.075$ near $y=1.2$ was used, improving the resolution near the wave crest. The time step was chosen $\Delta t=0.025$.

Figure 6.1 shows the computed velocity field and free-surface configuration for a third order solitary wave (Fenton, 1972), at the time instants $t=0$ and $t=15$. Figure 6.2 illustrates the vertical velocity contours for the same wave at time instants $t=10$ and $t=15$. These contour plots of vertical velocity show very clearly the position of the crest and the deformations suffered by the wave during the propagation. These deformations were a slight steepening of the front face, and small dispersive losses.

Table 6.1 shows the computed crest position x_{crest} , its theoretical value x_0+ct, and the computed wave amplitude H, for first, second and third order waves, at several time instants. The theoretical celerity was calculated using the dispersion relations appropriate for each order of approximation. For the first-order theory, the computed wave became higher and steeper and showed pronounced dispersive losses, but the computed celerity coincided with the theoretical value to within four significant digits. For the second order theory, the computed celerity was 4.2% higher than the theoretical value and the amplitude remained equal to its initial value, but the wave became progressively more asymmetrical (for $t=15$ it had a nearly vertical front). For the third-order wave, the computed celerity was 2.0% higher than the theoretical value, the amplitude after 15 time units was equal to the initial value, and the deformations and dispersive losses were smaller than in the previous cases.

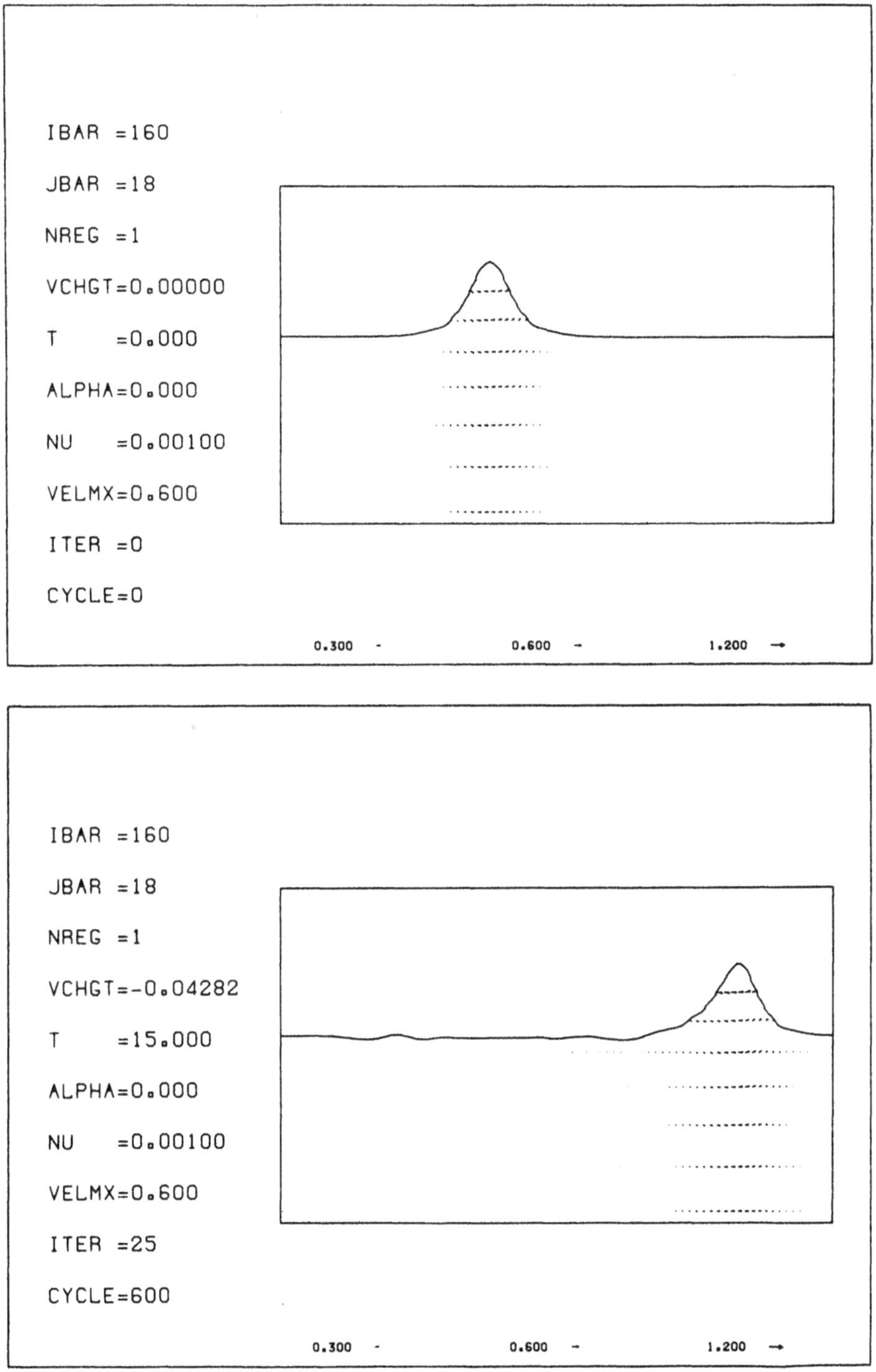

Figure 6.1 : *Velocity field and free-surface configuration for a third order solitary wave, at time instants* $t=0$ *(top) and* $t=15$ *(bottom). Vertical exaggeration* 1/0.0728 .

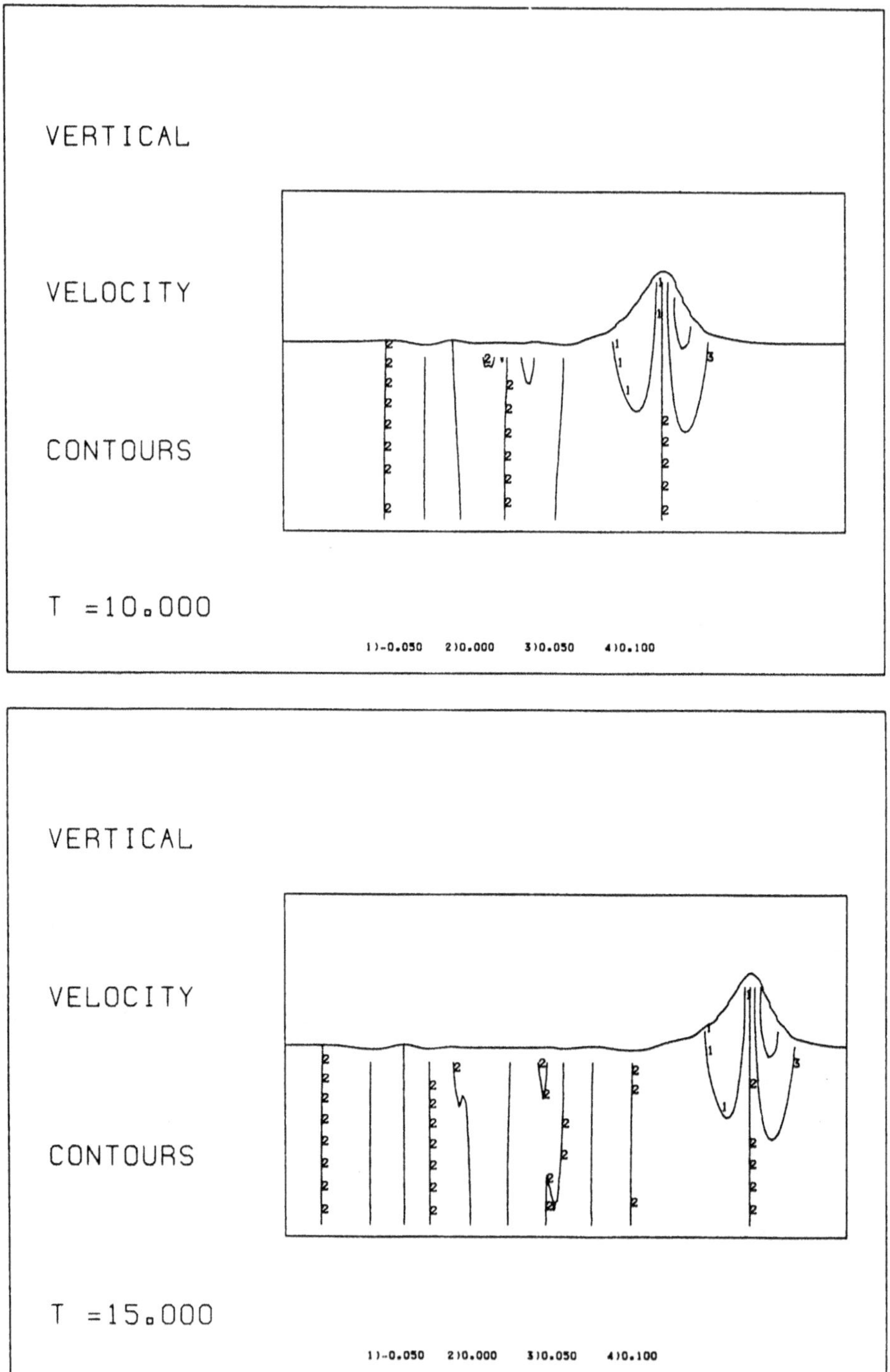

Figure 6.2 : *Vertical velocity contours for a third order solitary wave, at time instants* $t=10$ *(top) and* $t=15$ *(bottom). Vertical exaggeration* 1/0.0728.

Table 6.1 : Computed and theoretical crest position, and computed amplitude, for first, second, and third order solitary waves.

	t	x_{crest}	x_0+ct	H
1st	0	15.00	15.00	0.40
order	5	20.91	21.00	0.45
theory	10	27.00	27.00	0.46
2nd	0	15.00	15.00	0.40
order	5	21.07	20.88	0.40
theory	10	27.25	26.76	0.42
3rd	0	15.00	15.00	0.40
order	5	20.98	20.90	0.38
theory	10	26.66	26.79	0.38
	15	32.38	32.69	0.40

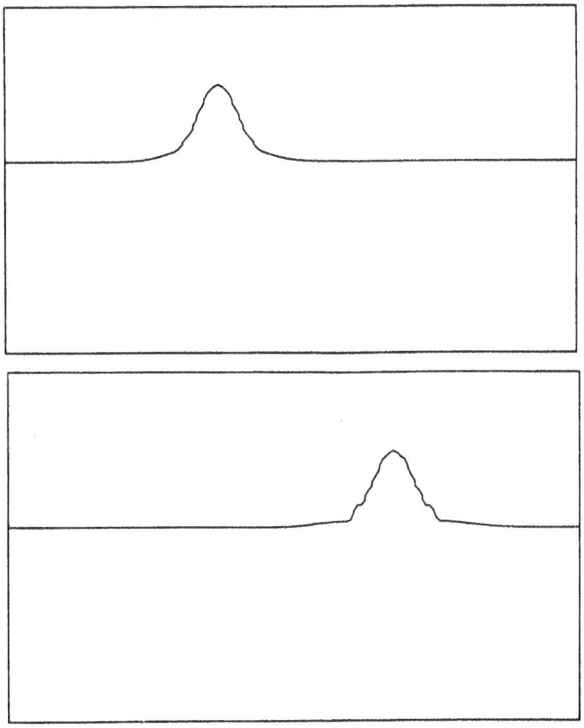

Figure 6.3 : *Free-surface profiles obtained in the free convection test for* $t=0$ *(top) and* $t=10$ *(bottom). Vertical exaggeration* 1/0.0728 .

To determine the influence of the F-convection algorithm in the observed forward tilting and dispersive losses, a test of free convection was performed. A third order wave profile was generated as before, but the velocity field of the wave was substituted by a uniform translation velocity of 1.2 . The acceleration of gravity and the viscosity were both chosen zero, so that the acceleration should also be zero in all mesh cells. Figure 6.3 shows the free-surface profiles for $t = 0$ and $t = 10$. There were small ripples on both sides of the crest, but these remained bounded and the wave crest maintained a symmetric form. Hence the deformations of the computed waves were not due to the F-convection algorithm.

In summary, it can be concluded that the model represents accurately the propagation of highly nonlinear waves. The deformations of the computed waves were due to the approximate initial and free-surface boundary conditions, and to the truncation errors. The best results were obtained using the third order theory. This indicates that the order of the wave theory has an important influence in the quality of the simulations.

6.3 Collision between solitary waves.

In contrast with most previous studies, this work is entirely concentrated on highly nonlinear waves and head-on collisions, instead of approximate models and overtaking collisions. The head-on collisions between solitary waves is a simplified model for the interaction between incident and reflected waves in a surf-zone. The collision problem is completely determined by the two dimensionless wave heights. Nearly equal dimensionless heights correspond to highly reflective conditions, whereas different dimensionless heights correspond to most real situations where dissipation is significant.

This type of interaction has been studied analytically (Byatt-Smith, 1971; Oikawa & Yajima, 1973), numerically (Fenton & Rienecker, 1982), and experimentally (Maxworthy, 1976). During the collision, the vertical acceleration of the fluid particles is large and approximate theories are no longer valid. After colliding, the original waves are deformed (Maxworthy, 1976; Fenton & Rienecker, 1982). Irregularity is introduced by the generation of secondary waves, due to dispersive losses. The amplitude and celerity of the original waves is changed. Furthermore, the original waves suffer a phase shift which some simplified theories fail to predict.

To evaluate the capability of the present technique for describing these complexities of wave-wave interactions, head-on collisions between solitary waves of equal and different amplitudes were simulated using 2D-HYDROTUR. For the case of waves of equal amplitude, the qualitative aspects of the solution and the computed phase shifts were compared with Maxworthy's (1976) experiments. For the case of waves of different amplitudes, the results showed that the waves see each other as a moving disturbance, their form being degraded by the collision.

6.3.1 Waves of equal amplitude.

Maxworthy (1976) performed experiments on wave-wave collisions and wave reflection by a vertical wall. He measured the maximum fluid height during the interaction and the phase shift suffered by the waves, and found that the computed phase shift is of $\mathcal{O}(D)$ for all dimensionless heights ϵ_w , whereas approximate theories predict that the phase shift increases with ϵ_w or is of the wrong sign (figure 6.4). After colliding, the waves are steeper and faster. This, too, cannot be explained using low order approximations. The differences between theory and experiment are due to the strong vertical accelerations during the interaction. To confirm these findings, and describe how the collision changes the interacting waves, this problem was investigated using 2D-HYDROTUR.

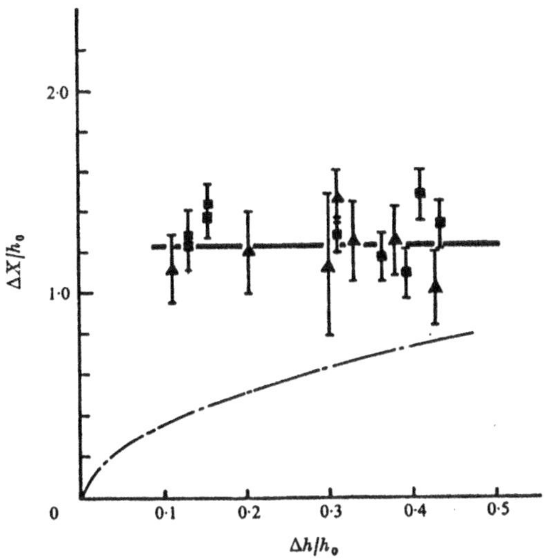

Figure 6.4 : *Magnitude of the dimensionless phase shift vs.* ϵ_w *for several values of water depth, and for wave-wave interaction (triangles) and wall reflection (squares). Dashed line : implicit result of Byatt-Smith* (1971) (*after Maxworthy,* 1976).

The physical conditions were chosen as close as possible to those of Maxworthy's (1976) experiments. The colliding waves were placed in a domain 5 m long and 0.2 m high with still water depth D=6.7 cm . The crests of the right- and left-running waves were positioned at $x=1$ m and $x=4$ m, respectively. The molecular viscosity was chosen $\nu=5\times10^{-4}$ m^2s^{-1} for stability reasons. The acceleration due to gravity was g=9.81 ms^{-2} downwards. The domain was discretized using a regular mesh of 200×20 cells with $\Delta x=2.5$ cm and $\Delta y=1$ cm . The time step was $\Delta t=5\times10^{-4}$ s. The boundary conditions were chosen free-slip on all mesh boundaries, and inviscid on the free-surface. Three numerical experiments were executed, for ϵ_w equal to 0.2, 0.35 and 0.45. Figures 6.5 to 6.8 show several stages of the collision between the two waves with $\epsilon_w=0.45$.

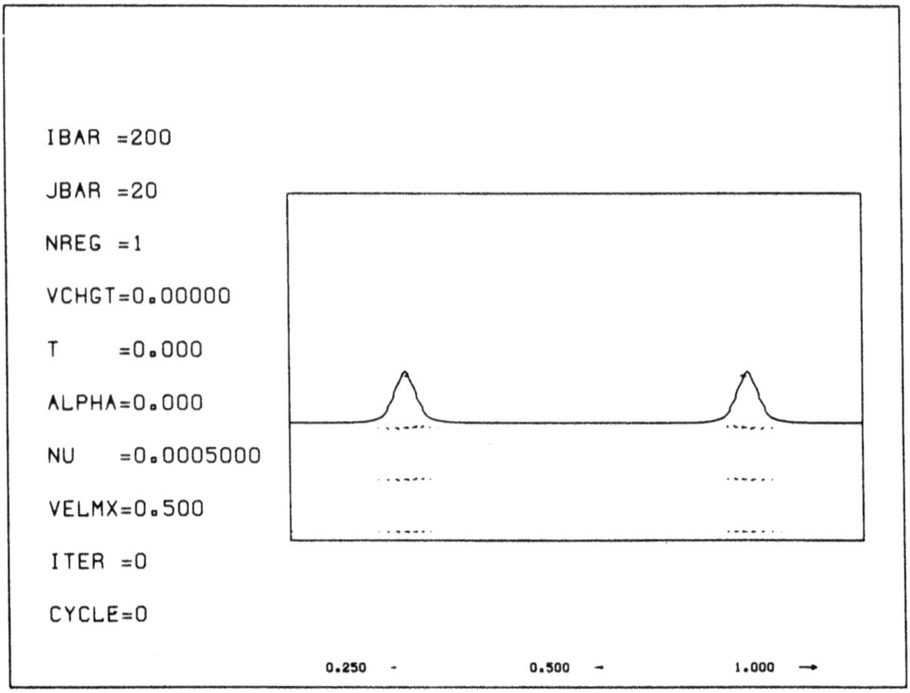

Figure 6.5 : *Collision between solitary waves of equal amplitude. Velocity field and free-surface configuration, $t = 0$ s . Vertical exaggeration* $1/6.74 \times 10^{-2}$.

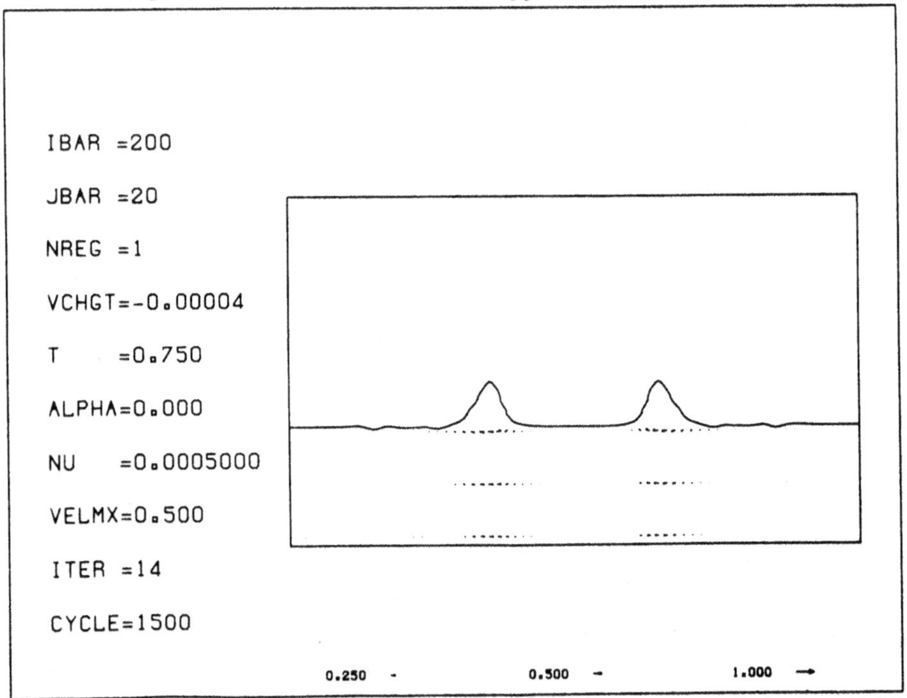

Figure 6.6 : *Collision between solitary waves of equal amplitude. Velocity field and free-surface configuration, $t = 0.75$ s . Vertical exaggeration* $1/6.74 \times 10^{-2}$.

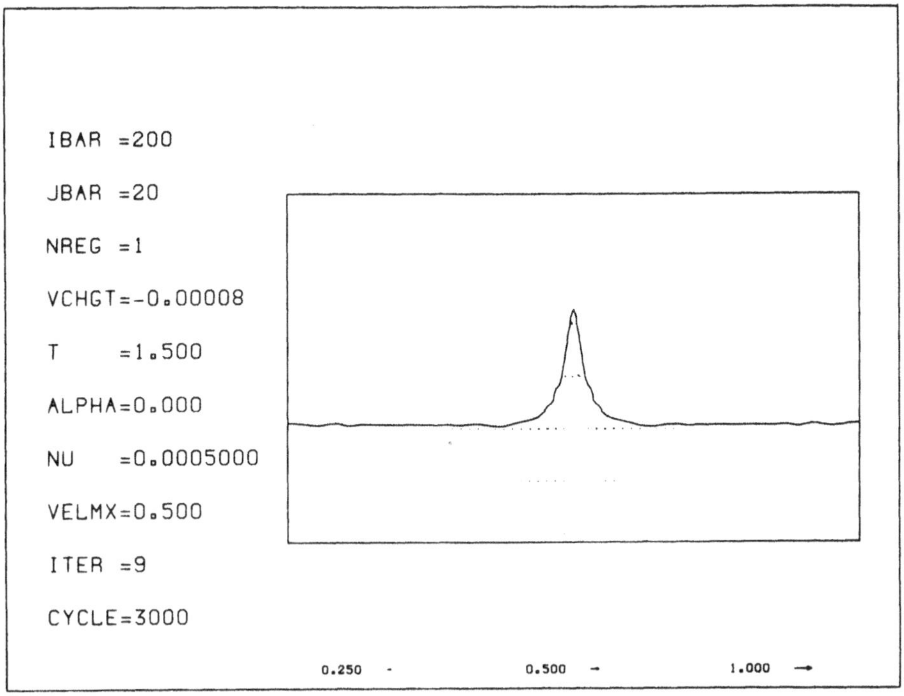

Figure 6.7 : *Collision between solitary waves of equal amplitude. Velocity field and free-surface configuration,* $t = 1.5$ *s . Vertical exaggeration* $1/6.74\times10^{-2}$.

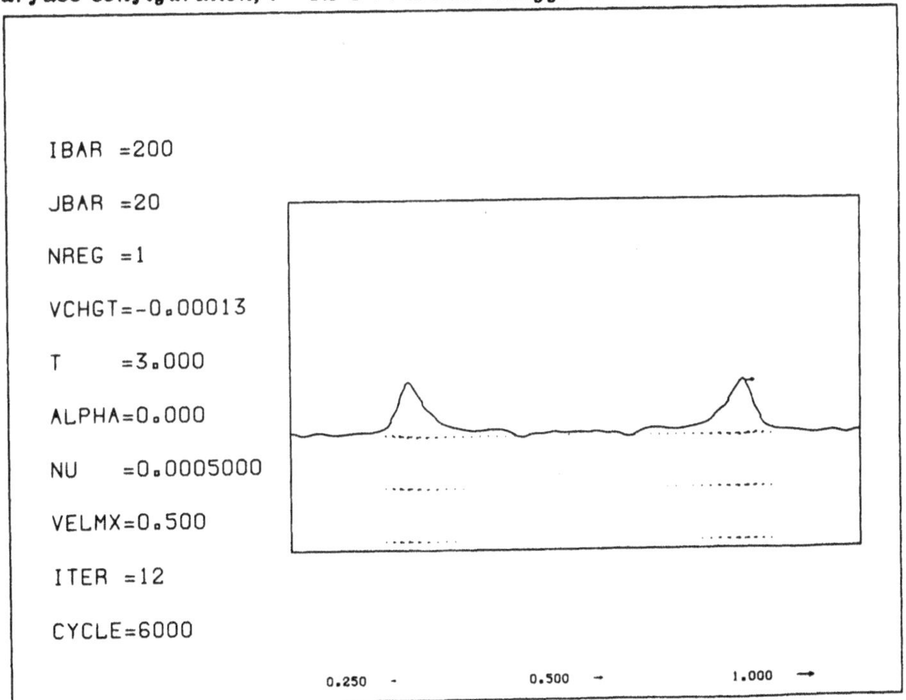

Figure 6.8 : *Collision between solitary waves of equal amplitude. Velocity field and free-surface configuration,* $t = 3.0$ *s . Vertical exaggeration* $1/6.74\times10^{-2}$.

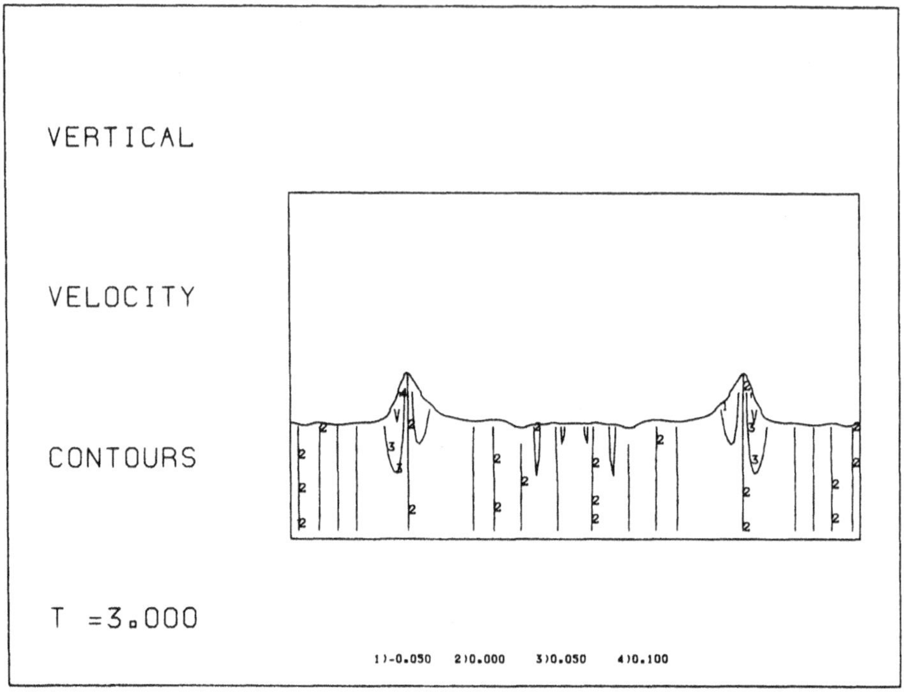

Figure 6.9 : *Collision between solitary waves of equal amplitude. Vertical velocity contours*, $t = 3.0$ *s* . *Vertical exaggeration* $1/6.74\times10^{-2}$.

Before the collision, the interacting waves were slightly lower than at the beginning. They were also slightly asymmetrical, each of them shedding small dispersive waves. These relatively unimportant deformations appeared to be due to simple propagation. During the collision, the waves merged into a single hump of fluid whose height was larger than $2H$, and whose width was of $\mathcal{O}(D)$.

After the interaction the waves were higher and steeper than before the collisions. Also, they were clearly tilted forwards. This is best shown in figure 6.9, which illustrates the vertical velocity contours for $t = 3$ *s* . The dispersive losses were also more pronounced than before the collision. Figure 6.9 also shows that the solution is perfectly symmetric about the center of mass of the two waves. This demonstrates the isotropy of the algorithm, a property that was also observed in all other solutions obtained in this set of simulations. In the simulations with ϵ_w=0.20 and ϵ_w=0.35 (not shown), the deformations (tilting and dispersive losses) were much smaller than for ϵ_w=0.45, and the waves emerged from the collision without significant changes in their structure.

The phase shift suffered by the waves and the celerity before and after the interaction were calculated from the crest positions for the left-running wave (the other wave would give identical results). Figures 6.10–6.12 show the successive crest positions *vs* the phase function for and undisturbed wave, for the three numerical experiments. Table 6.2

shows the dimensionless phase shift during the interaction, and the celerity before and after the collision. For $\epsilon_w = 0.20$ the computed phase shift was smaller than the value reported by Maxworthy, but closer to the experimental value than to the approximate analytical prediction. For $\epsilon_w = 0.35$ the computed phase shift was slightly higher than the experimental value (Maxworthy found $\Delta x_{crest}/D = 1.40$ for $\epsilon_w = 0.31$). For $\epsilon_w = 0.45$ the computed phase shift was inside the error limits reported by Maxworthy, hence much closer to the experimental values than the results of approximate theories.

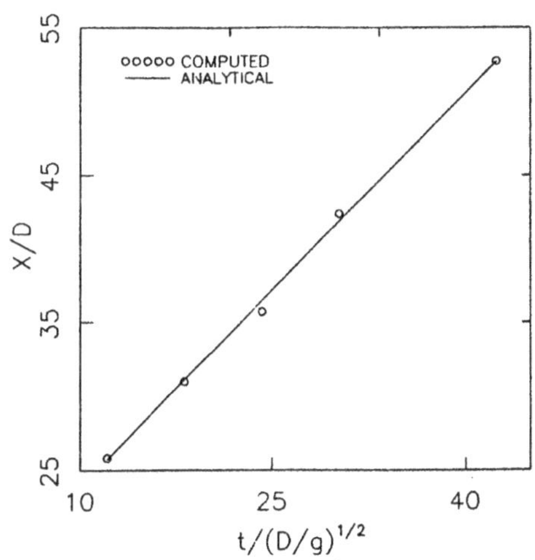

Figure 6.10 : *Dimensionless wave crest position vs. dimensionless time, for the computed solution (circles) and for an undisturbed wave (solid line),* $\epsilon_w = 0.20$.

In summary, it was shown that the numerical model can predict difficult aspects of wave-wave interaction, with moderate spatial resolution and computer cost. Some features of the problem were dependent on ϵ_w, while others were not. The steepening and deformation of the original waves, and the appearance of secondary waves, was clearly dependent on ϵ_w. Both increased with ϵ_w . Apparently, these effects are not significant for moderate dimensionless amplitudes (say $\epsilon_w < 0.30$). The increase of the wave celerity after the interaction varied between 2.4% and 3.0% , and thus seems to be independent on ϵ_w. This supposition needs further confirmation. It cannot be concluded that the phase shift is constant for all ϵ_w. Both Maxworthy's experiments and the present simulations seem to indicate that the phase shift is maximum for $0.31 < \epsilon_w < 0.35$.

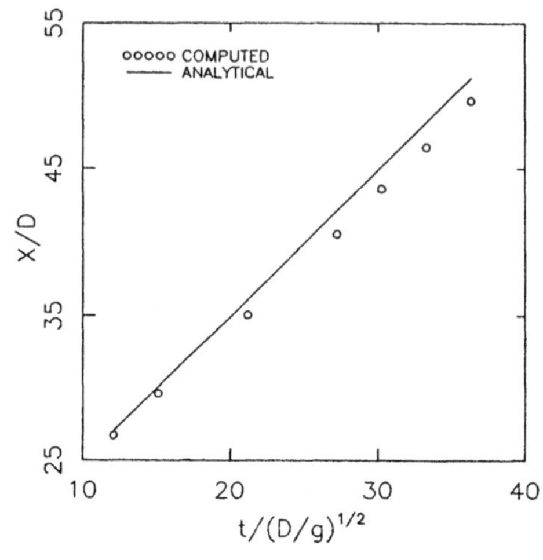

Figure 6.11 : *Dimensionless wave crest position vs. dimensionless time, for the computed solution (circles) and for an undisturbed wave (solid line),* $\epsilon_w = 0.35$.

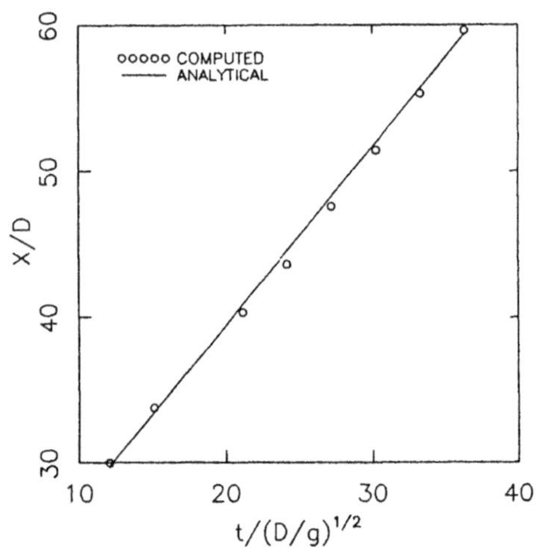

Figure 6.12 : *Dimensionless wave crest position vs. dimensionless time, for the computed solution (circles) and for an undisturbed wave (solid line),* $\epsilon_w = 0.45$.

Table 6.2 : Computed dimensionless phase shift $\Delta x_{crest}/D$, and (dimensional) celerities before and after the interaction for the three numerical experiments.

ϵ_w	Phase Shift	Celerity before interaction	Celerity after interaction
0.20	0.78	0.90 ms^{-1}	0.92 ms^{-1}
0.35	1.50	0.97 ms^{-1}	1.00 ms^{-1}
0.45	0.91	1.25 ms^{-1}	1.28 ms^{-1}

6.3.2 Waves of different amplitude.

The collision between waves of different amplitude is more relevant for the hydrodynamics of pre-breaking waves than the case of waves of equal amplitude, because in most surf-zones dissipation is high, and the reflected waves are less energetic than the incident waves. It would be interesting to know if the properties of head-on collisions between waves of different amplitude are similar to those of equal-amplitude waves (negative phase shift, forward tilting, dispersive losses), and if so, what is the influence of the difference between the two wave heights in the outcome of the interaction.

The collision problem is now determined by two parameters, the dimensionless wave heights. It is known that the form of the original waves is degraded (if they are sufficiently high), the bigger wave is tilted forwards, and small dispersive waves are generated (Schwartz & Fenton, 1982; Fenton & Rienecker, 1982). Also, the phase shifts resulting from head-on collisions are different from those resulting from overtaking collisions between waves with the same amplitudes.

In this section, a numerical experiment of collision between two waves, one with ϵ_w=0.25 and the other with ϵ_w=0.5, is described. As in § 6.2 , time and length units were chosen so that $D=1$ and $g=1$. The two solitary waves were placed in a domain 90 units long and 2 units high. A right-running wave with $H=0.25$ units was initially placed at $x=25$, and a left-running wave with $H=0.50$ units was positioned at $x=60$. The water viscosity was set to $\nu=10^{-2}$ for stability reasons. The domain was discretized using a mesh of 250×20 cells, with $\Delta x=0.36$ and $\Delta y=0.10$.

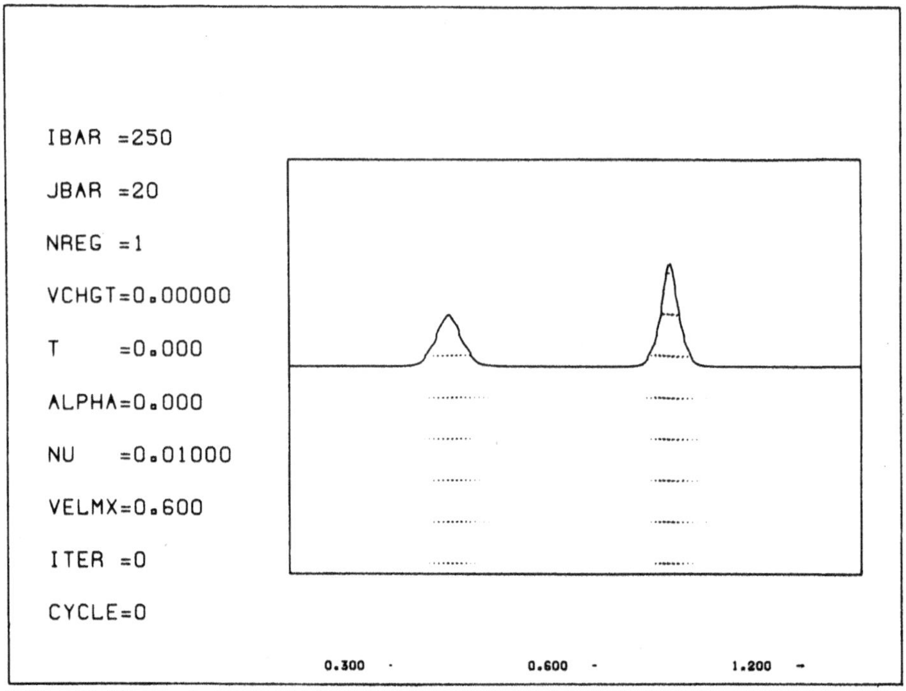

Figure 6.13 : *Collision between solitary waves of different amplitudes. Velocity field and free-surface configuration, t*=0. *Vertical exaggeration* 1/0.03 .

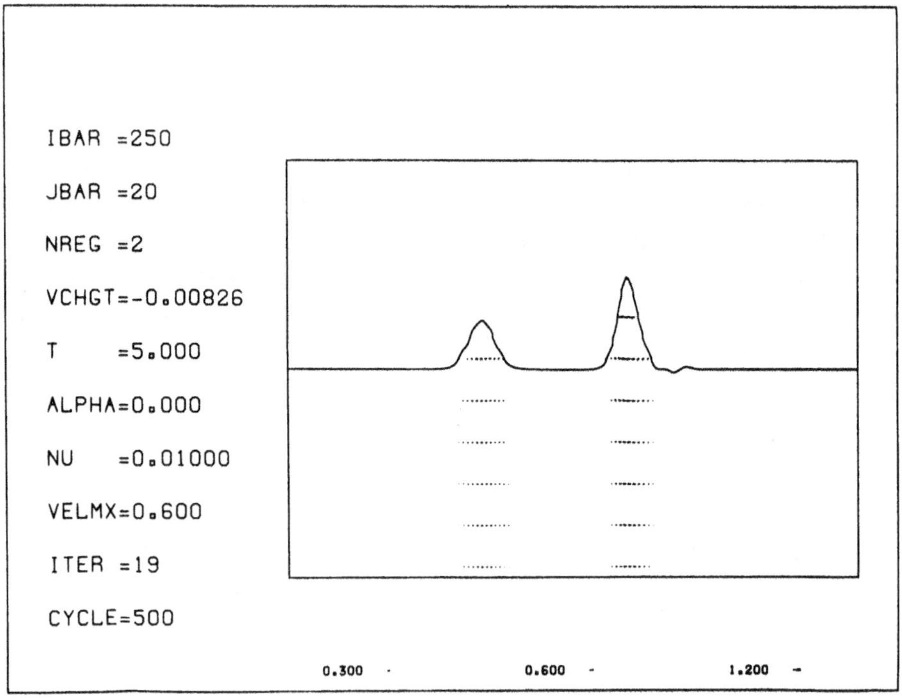

Figure 6.14 : *Collision between solitary waves of different amplitudes. Velocity field and free-surface configuration, t*=5. *Vertical exaggeration* 1/0.03 .

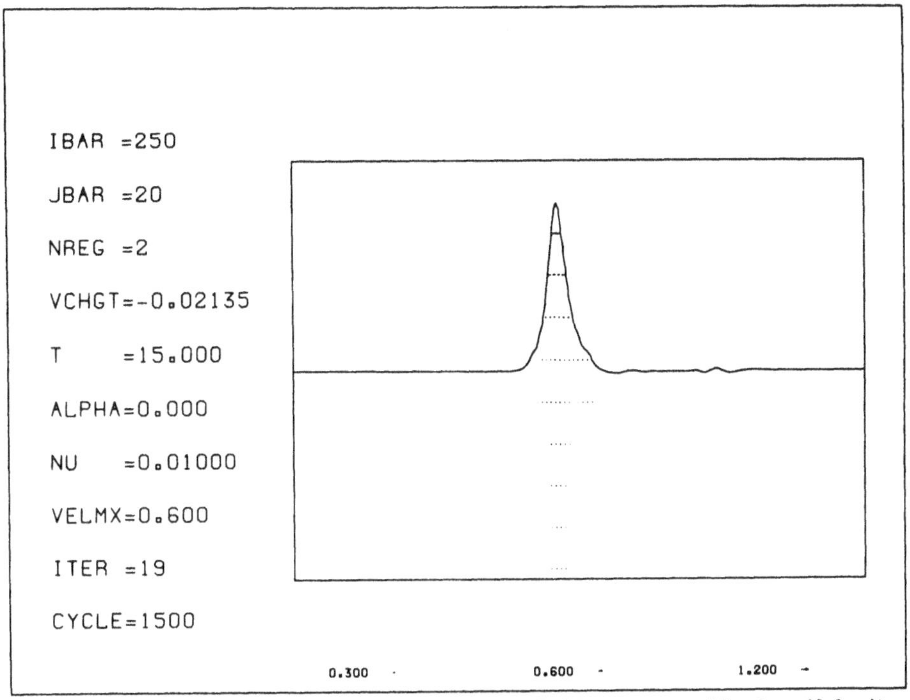

Figure 6.15 : ***Collision between solitary waves of different amplitudes. Velocity field and free-surface configuration, t=15. Vertical exaggeration*** **1/0.03 .**

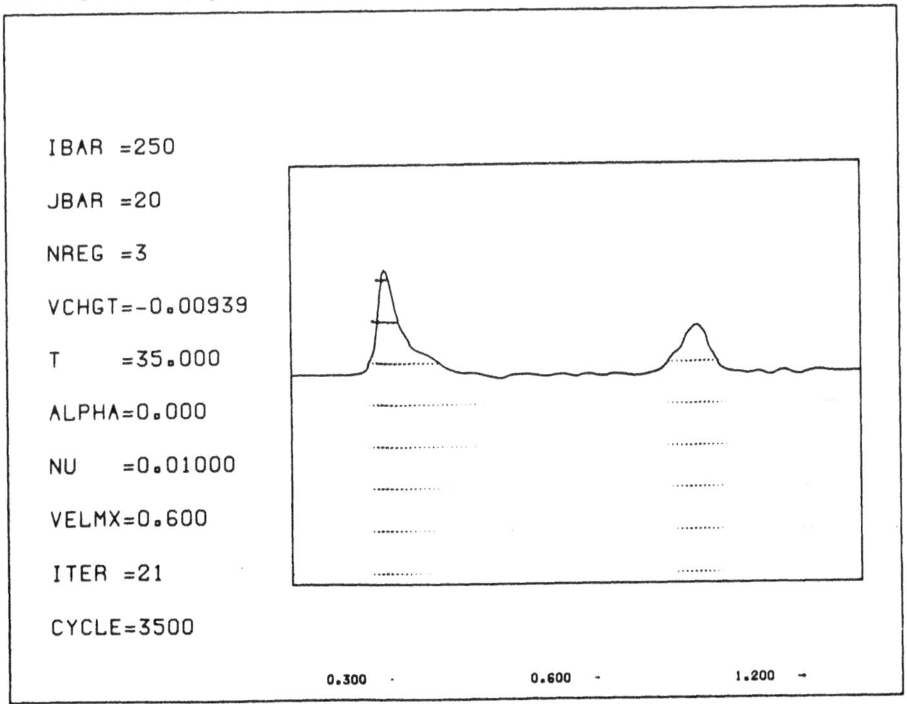

Figure 6.16 : ***Collision between solitary waves of different amplitude. Velocity field and free-surface configuration, t=35. Vertical exaggeration*** **1/0.03 .**

Figures 6.13–6.16 show the instantaneous velocity field and free-surface configuration of the wave system for several successive instants. Before the interaction, the deformation of the waves was small. The smaller wave maintained its original shape and amplitude, without noticeable dispersive losses. The bigger wave was slightly deformed, showing dispersive waves behind the crest. During the collision, a large hump of fluid was formed near the center of the domain. Because the two waves were not equal, large horizontal velocities were found close to the fluid peak. Also, new dispersive waves appeared, indicating a degradation of the incident waves during the collision. After colliding, both waves were tilted forwards. The bigger wave was then much steeper than before the interaction. The smaller wave was comparatively less degraded. Therefore, it can be concluded that the deformations arising from head-on solitary wave collisions are stronger for waves of different amplitudes than for waves of equal amplitude.

Table 6.3 shows the computed wave heights and crest positions for the collision between solitary waves of different amplitudes.

Table 6.3 : Computed wave heights and crest positions for the colliding solitary waves of different amplitudes.

	$\epsilon_w = 0.25$		$\epsilon_w = 0.50$	
t	x_{crest}	H	x_{crest}	H
0	25.0	0.25	60.0	0.50
5	30.7	0.23	53.8	0.50
25	52.2	0.21	28.5	0.50
35	63.8	0.22	15.3	0.47

The bigger wave suffered a dimensionless phase shift of −0.4 , while the smaller wave suffered a dimensionless phase shift of −0.9 . After the collision, the celerity of the bigger wave increased 5.6% , and the celerity of the smaller wave increased 2.6 % . The most important feature of the interaction was the deformation of the bigger wave, which developed a "bump" in its back face. This type of deformation is observed in solitary wave shoaling (Mei, 1983), and eventually leads to the fission of the original wave. Such a qualitative similarity suggests that the bigger wave sees the smaller wave as a moving obstacle, acting for a short period of time. The unexpected result that both wave heights decrease was also reported, although for different values of ϵ_w, by Fenton & Rienecker (1982) . The decrease of height of both waves can in part be attributed to the dispersive losses.

Some interesting conclusions concerning the hydrodynamics of pre-breaking waves were obtained. It was shown that the sequence *disturbance* ⇒ *evolution* ⇒ *breaking* can be started by a wave-wave collision. Wave irregularity can in part be explained in terms of dispersive waves that are generated by the interaction. Because the collision leaves the waves in a condition closer to breaking (they become steeper and with higher horizontal velocities near the wave crest), wave-wave collisions may be a factor influencing the irregularity of the breaking point. **However, it should be noted that in real surf-zones, wave interaction and wave irregularity are much more complicated than suggested by the simple model here considered**. The model predicted the properties of strong wave interactions with fair accuracy. For the case of waves of equal amplitude, it was possible to reproduce all qualitative features of the interaction problem with low spatial resolution and computer cost (each simulation took about 1 hour of CPU time on a VAX 8600 computer operating at 500-750 KFLOPS).

6.4 Simulation of undular, transitional and turbulent hydraulic jumps.

There are several reasons for including the hydraulic jump in the present investigation. It is a well studied type of flow (Rajaratnam, 1967; Resh & Leutheusser, 1972; Madsen & Svendsen, 1983) which allows simple and accurate checks of mass and momentum conservation, and can be generated using simple initial and boundary conditions. Also, the turbulent hydraulic jump is a useful model for describing breaking waves in the inner region. However, the most important reason is that the jump's behavior can be either undular (dispersive) or turbulent (dissipative), depending on the Froude number. Since the essence of surf-zone wave hydrodynamics is the transition from dispersive to dissipative regime, it is natural to ask if the nature of the transition and the characteristics of the resulting turbulence field can be described using the hydraulic jump as the basic model.

Previous experimental and numerical investigations on the hydraulic jump left several questions unanswered. The experimental and numerical estimates for the critical depth ratio obtained using low order nonlinear models are very different (§ 3.5.2). Numerical simulations of the detailed flow structure in turbulent jumps are not available.

In the present investigation, moving hydraulic jumps were generated by pushing a horizontal uniform stream into a rigid vertical wall. Different qualitative behavior was produced by varying the Froude number of the incident stream. Figure 6.17 schematizes the generation of a moving jump.

The discrete mass and momentum conservation equations in a fixed reference frame are

$$u_2 = (1 - \frac{h_1}{h_2})C \tag{6.1}$$

$$C = [g\frac{h_2}{h_1}(\frac{h_2 + h_1}{2})]^{1/2} \tag{6.2}$$

These equations can be solved for h_1 and C given g, u_2 and h_2 . The reason why moving jumps were used instead of steady jumps is that the *computed* h_1 provides a direct check of momentum conservation. If steady jumps were generated by specifying velocity and water depth on both sides of the jump, momentum conservation would be enforced automatically *via* the boundary conditions. Moving jumps can be rendered stationary by adding the jump celerity to the velocity field, except for bottom boundary layer dynamics. However, experimental evidence (confirmed by the present study) shows that turbulence generation in the bottom boundary layer is much smaller than turbulence generation by breaking (Resh & Leutheusser, 1972; Madsen & Svendsen, 1983).

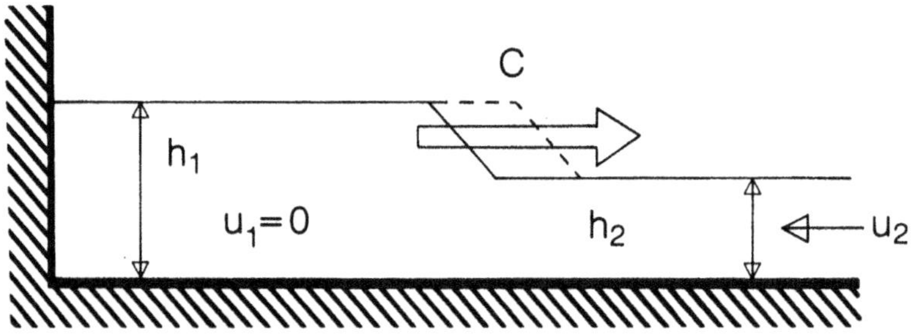

Figure 6.17 : *Schematization of the moving hydraulic jump.*

This section is organized as follows. In § 6.4.1 , the development of an undular bore is calculated, and the characteristics of the jump waves are compared with experimental results. In § 6.4.2 the critical depth ratio is determined and the nature of the transition regime is discussed. Also, the discrepancy between the previous experimental and numerical estimates of this parameter is analyzed. Finally, in § 6.4.3 the full capabilities of 2D-HYDROTUR are used to compute the evolution of two turbulent (strong) jumps with depth ratios 2.17 and 3.52. The qualitative differences between these two jumps is discussed. The relevant turbulent field quantities are made dimensionless using the appropriate velocity scale $u_2 - C$ (rendering the jump steady), and their distributions are compared with the experiments of Resh & Leutheusser (1972).

6.4.1 Undular jump.

In this section, the properties of an undular jump, computed using 2D-HYDROTUR, will be compared with previous experimental (Favre, 1935) and numerical (Abbott & Rodenhuis, 1972) results. The wavelength of jump waves is exceptionally sensitive to variations in incident energy and momentum fluxes. Numerical simulations prior to Abbott & Rodenhuis (1972) provided jump waves with wavelengths differing from the experimentally observed values by 30 to 300% . Abbott & Rodenhuis (1972) computed the evolution of undular jumps using a Boussinesq-type system operated over a range of values of Δt , and extrapolated the computed wavelength and crest positions for $\Delta t = 0$, thus filtering out the numerical perturbations (truncation errors). The extrapolated results for $t \rightarrow \infty$ yielded wavelengths close to Favre's to within 3% .

The physical parameters of the undular jump problem were chosen as close as possible to those of Favre's experiment nr. 22 , given the computer resources available. The physical domain was 6 m long and 0.2 m high, with a stream depth of 0.1075 m . The domain was discretized using a regular mesh of 200×20 cells with $\Delta x = 3\ cm$ and $\Delta y = 1\ cm$. A uniform stream of fluid was generated by imposing an initial horizontal velocity of $-0.1386\ ms^{-1}$. All boundaries were specified free-slip, except the right boundary, which was left open. This allowed a continuous input of fluid, preventing any waves from being generated at the inflow boundary and at the same time providing another test for the open boundary treatment. With these data, the sequent depth ratio is $h_1/h_2 = 1.14$, the supercritical Froude number is $\mathbf{F} = |u_2 - C|/(gh_2)^{1/2} = 1.104$, and the jump celerity is $C = 0.995\ ms^{-1}$. The acceleration due to gravity was g=9.81 ms^{-2} downwards. The molecular viscosity was chosen $\nu = 10^{-4}\ m^2 s^{-1}$ for stability reasons, and the turbulence model was not activated. The time step was $\Delta t = 10^{-2}\ s$.

Figures 6.18 to 6.20 illustrate the propagation of the undular jump for the times $t = 1\ s$, $t = 2\ s$ and $t = 4\ s$, by showing the instantaneous velocity field and free-surface configuration of the fluid. It was observed that the number of waves present is proportional to the time since the jump was started. This result agrees with Favre's observations and confirms the hypothesis that the energy is carried away from the jump's front by the wave group (Lemoine, 1948; Benjamin & Lighthill, 1954).

The computed velocity field distribution was well described by the shallow water approximation (arising from Rayleigh-Boussinesq expansions—§ 3.3.1): the horizontal velocity was nearly depth-uniform, and the vertical velocity increased linearly from the bottom to the free-surface, in any vertical section of the jump. The development of the jump waves was progressive, and each of them approached a nearly-symmetrical form. These aspects of the solution are best seen in figure 6.21 , which shows a contour plot of vertical velocity. This vertical velocity plot shows 5 clearly developed wave crests.

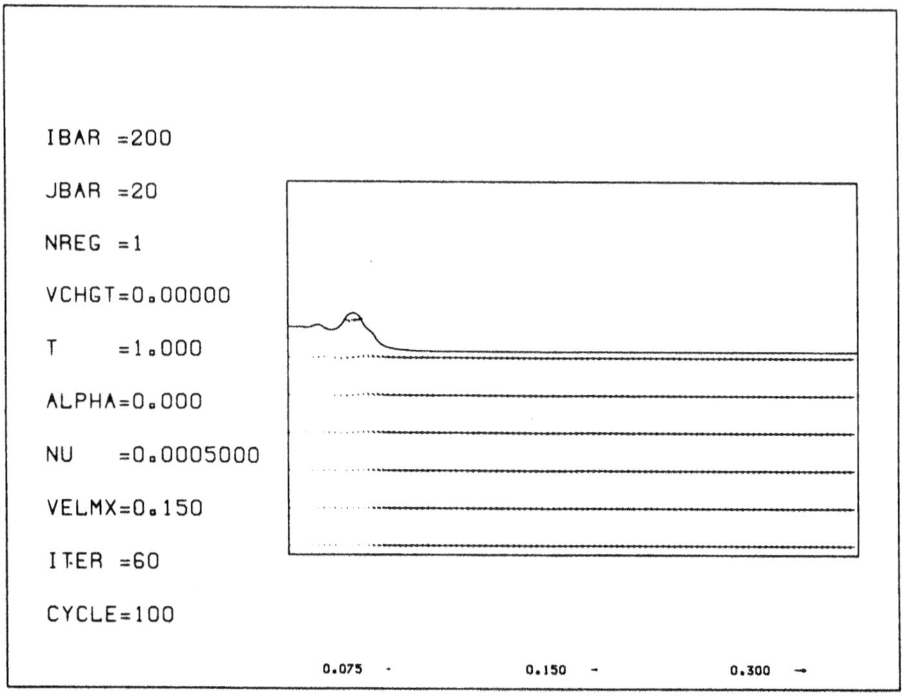

Figure 6.18 : *Velocity field and free-surface configuration for the undular jump,* $t = 1.0$ *s . Depth ratio* 1.14 ; $\mathbf{F} = 1.104$; *vertical exaggeration* 1/0.05

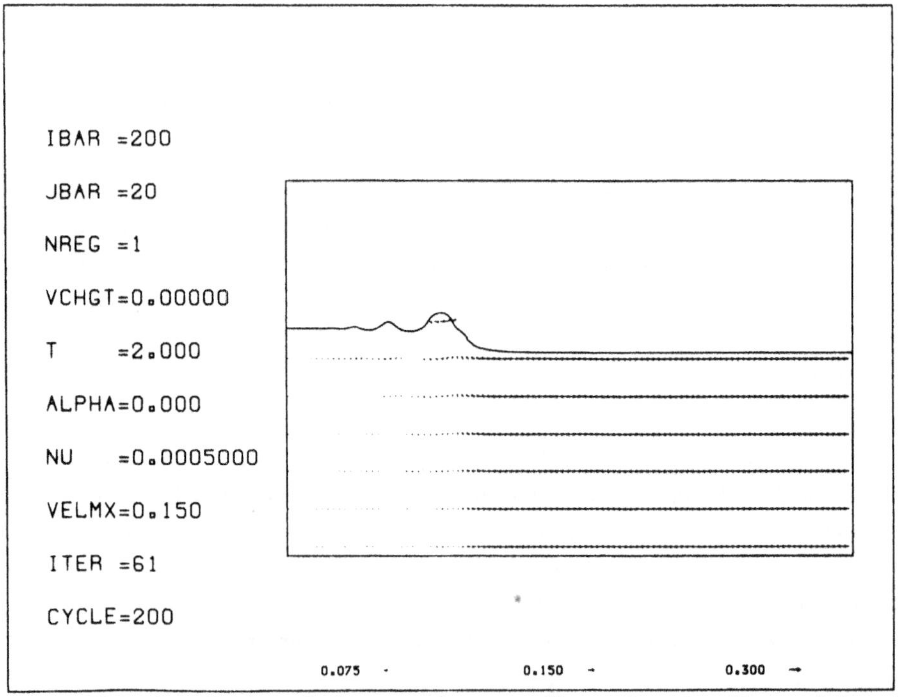

Figure 6.19 : *Velocity field and free-surface configuration for the undular jump,* $t = 2.0$ *s . Depth ratio* 1.14 ; $\mathbf{F} = 1.104$; *vertical exaggeration* 1/0.05 .

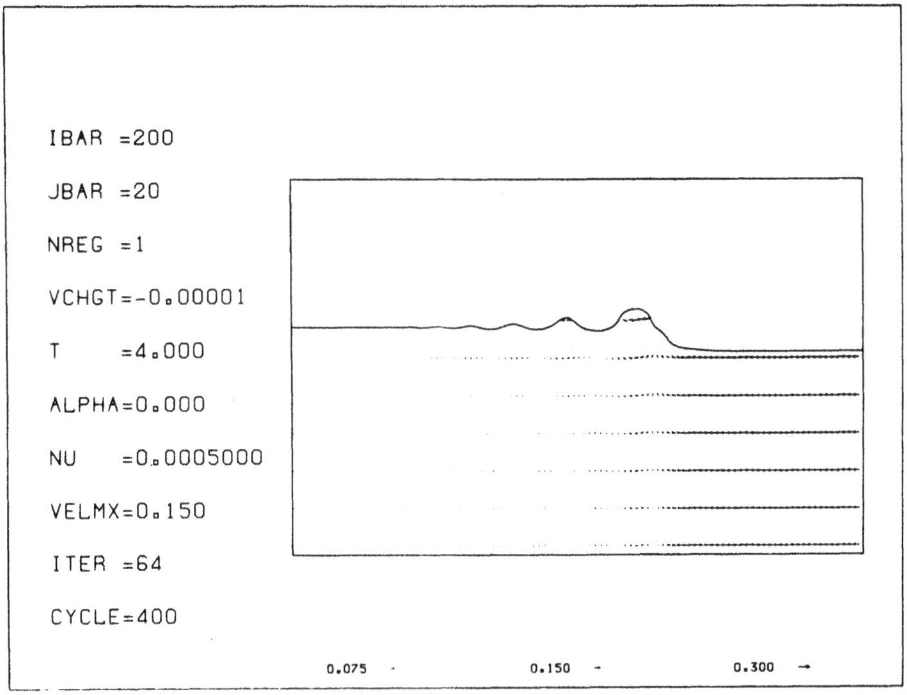

Figure 6.20 : *Velocity field and free-surface configuration for the undular jump,* $t = 4.0$ *s* . *Depth ratio* 1.14 ; $\mathbf{F} = 1.104$; *vertical exaggeration* 1/0.05 .

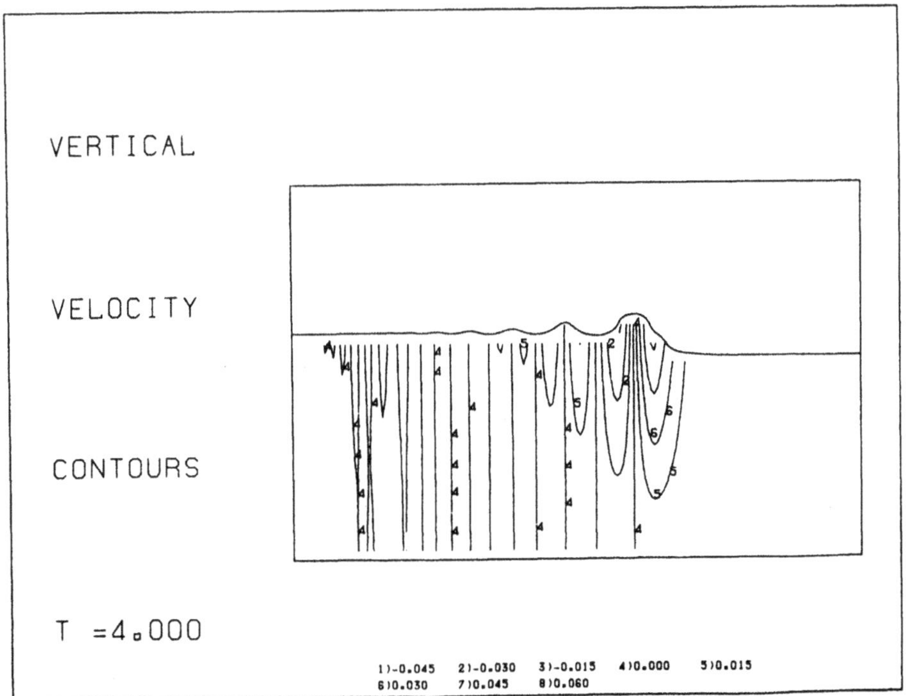

Figure 6.21 : *Vertical velocity contours for the undular jump,* $t = 4.0$ *s* . *Depth ratio* 1.14 ; $\mathbf{F} = 1.104$; *vertical exaggeration* 1/0.05 .

The wavelength of the jump waves was computed from the successive positions of the two leading crests, and plotted against $1/t$ to find the asymptotic value λ_∞ (figure 6.22). Using a trial and error method, it was found that the expression

$$\lambda = exp(0.0578 - 1.53778/t + 0.59862/t^2)$$

provided an extremely good fit to the computed wavelengths.

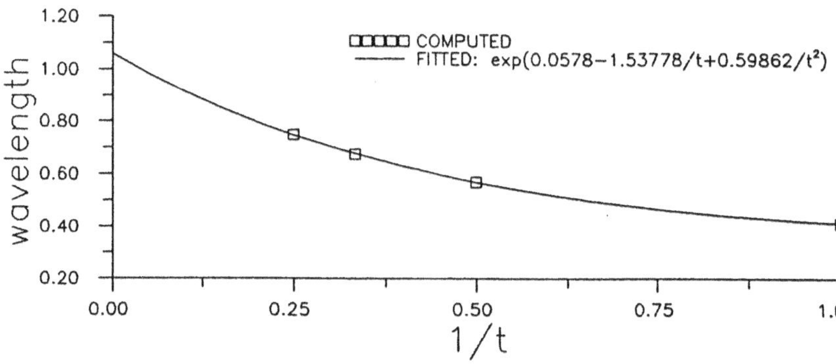

Figure 6.22 : *Computed and fitted wavelength for jump waves.*

The computed asymptotic wavelength λ_∞ was 1.06 m, whereas the wavelength observed by Favre is 1.01 m and the numerical computation of Abbott & Rodenhuis (1972) yielded 1.04 m. The linear approximation of Lemoine (1948) gives the estimate 0.97 m, which is clearly inferior to the numerical and experimental results. The amplitude of the jump waves is less sensitive than the wavelengths. The asymptotic elevation of the first wave crest obtained in the present work was 0.135 m, in agreement with Favre's result. In contrast, Abbott & Rodenhuis (1972) give 0.140 m. The cause of this discrepancy is not clear. The elevation of the first crest estimated using Lemoine's (1948) theory is 0.132 m, which is in good agreement with the present numerical results and with the experimental value. The computed fluid height at the wall after 4 seconds was 0.1218 m, with an error of 0.65% to the theoretical value 0.1225 m.

6.4.2 Transitional jump.

After characterizing the development of a moving jump in dispersive regime, the following questions were considered :

- What is the critical depth ratio for breaking, calculated with the present model ?

- What are the distinguishing features of the velocity field and free-surface configuration close to the critical conditions ?

The critical depth ratio for breaking is another sensitive jump characteristic. Abbott & Rodenhuis (1972) explained the observed discrepancy between the computed and the experimentally observed critical depth ratio in terms of real fluid effects (not present in their formulation), such as viscosity, initial conditions, physical conditions of the flume, etc. Accordingly, they argue that all ideal-fluid jumps break, regardless of the depth ratio. Abbott (1979) interpreted the formation of a turbulent jump in terms of breaking of the jump waves, which become higher and more unstable as the depth ratio increases.

The depth ratio can be increased by imposing a higher Froude number. For a given stream depth, this can be done by either increasing the stream velocity, or by decreasing the acceleration of gravity. In the present study the latter alternative was chosen, this allowing the simulations to be done with smaller velocities (hence with less stability problems). First, time and length units where chosen so that $D=1$ and $g=1$. The domain, 30 units long and 3.2 units high, was discretized using a regular uniform mesh of 80×16 cells. The viscosity was chosen $\nu=10^{-3}$ for stability reasons.

To investigate the nature of the transition, two jumps were simulated. The first jump, with $h_1/h_2=1.28$, was generated by imposing an initial stream velocity of -0.265 . The celerity was $C=0.944$, and the supercritical Froude number was $\mathbf{F}=1.21$. The second jump, with $h_1/h_2=1.30$, was generated by imposing an initial stream velocity of -0.283 , and the supercritical Froude number was $\mathbf{F}=1.22$. In both simulations all boundaries except the right boundary were chosen free-slip. The right boundary was specified open, following the procedure of the previous section.

Figure 6.23 shows the computed velocity field and free-surface configuration for the moving jump with sequent ratio 1.28 , for $t=12.5$, $t=22.5$, and $t=25$.

It was observed that the velocity field structure is qualitatively different from that of an undular jump. The vertical velocity is not proportional to the distance from the bottom, in the body of the jump. The horizontal velocity is not nearly-uniform in y , for any vertical section behind the front of the jump. The prominent feature of this velocity field is a strong, localized convergence into a backward-directed jet close to the jump's propagating front. This jet is qualitatively similar to the plunging jet in spilling and plunging breaking waves (§ 3.2.3).

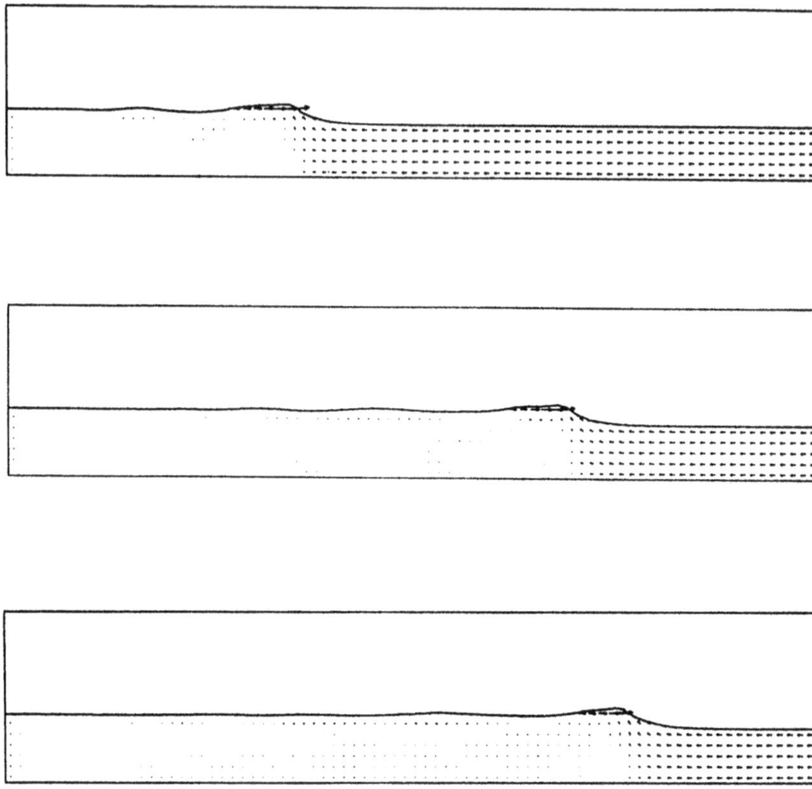

Figure 6.23 : *Velocity field and free-surface configuration for a moving jump with depth ratio* 1.28, *for* $t = 12.5$ (*top*), $t = 22.5$ (*middle*) *and* $t = 25$ (*bottom*). *Supercritical Froude number* : 1.21 .

It was also observed that the backward jet close to the front of the jump did not grow with time. Its configuration and strength at $t = 12.5$ and $t = 25$ were nearly identical. These two observations suggest the conclusion that the jump was transitional, but stable.

In the initial stage of the propagation, small undulations in the free-surface were observed behind the jump's front. Subsequent plots of the solution showed that these undulations tended to disappear, and played no relevant role in the behavior of the jump. The explanation for this fact was found in the structure of the velocity field. Since this latter did not show a pronounced undular character, there could be no strong undulations in the free-surface.

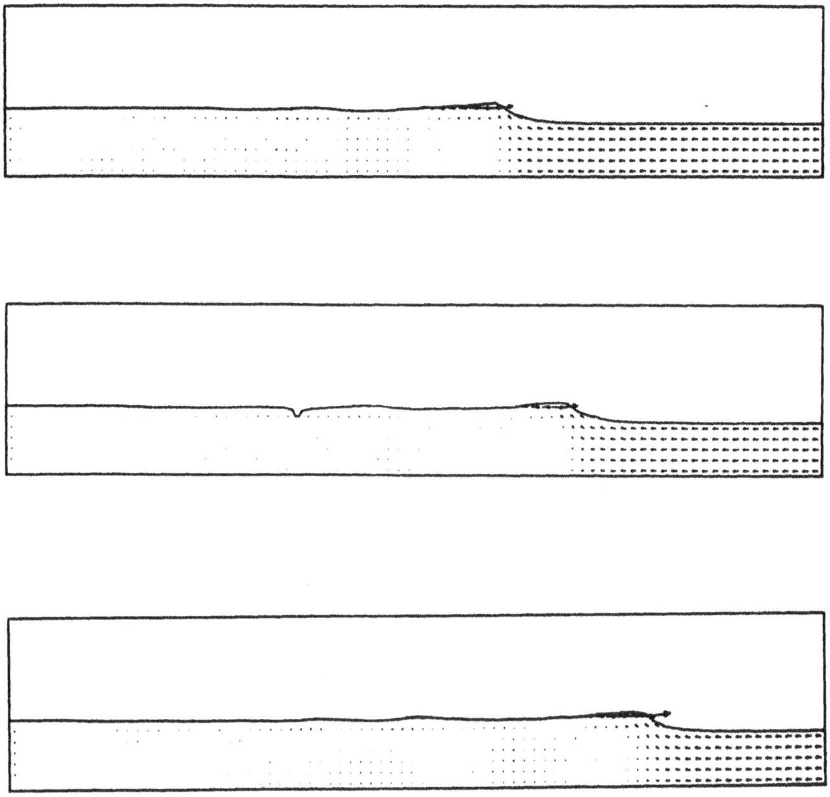

Figure 6.24 : *Velocity field and free-surface configuration for a moving jump with depth ratio* 1.30, *for* $t=20$ *(top)*, $t=22.5$ *(middle) and* $t=25$ *(bottom). Supercritical Froude number* : 1.22 .

Figure 6.24 shows results of the numerical simulation of a jump with depth ratio 1.30 ($\mathbf{F}=1.22$), for $t=20$, $t=22.5$ and $t=25$. The velocity field configuration looked similar to the previous one, but the backward ("plunging") jet grew rapidly after $t=22.5$ s . Between $t=22.5$ and $t=25$, the backward jet became stronger and more localized, and the front face of the jump steeper. This indicated that the jump was unstable.

Also, it was observed that the undulations of the free-surface (and of the velocity field) were almost totally absent. This suggests that the breaking of the hydraulic jump is not originated by the breaking of jump waves (Abbott, 1979), but by the intensification of the localized jet close to the jump's front. Recent numerical results by Teles da Silva & Peregrine (1990) also support this conclusion.

In summary, the results indicate that transition occurs for $1.28 < h_1/h_2 < 1.30$, or , equivalently, for $1.21 < \mathbf{F} < 1.22$, in agreement with the experimental value. It is remarkable that this critical parameter can be determined so accurately using very modest spatial resolution and computer resources (both transitional jump calculations were done in a small IBM-PC). The prominent feature of transitional jumps is the convergence of the velocity field into a plunging jet close to the jump front, resembling that of a pre-spilling (because it is very narrow) breaking wave (further evidence of this was presented by Teles da Silva & Peregrine, 1990). Therefore, it appears that this characteristic of the velocity field is common to all situations were the balance between nonlinearity and dispersivity is about to be broken, a finding which reinforces the theoretical interest of the hydraulic jump. It seems that the only difference between pre-plunging/spilling waves and transitional jumps is that in the former the disturbance (bottom topography) acts continuously, whereas in the latter the disturbance (wall) is impulsive. The failure of some previous attempts to determine the critical depth ratio was due to the nature of the approximations used in the numerical approaches. Close to critical conditions, the velocity field cannot be described in terms of shallow water approximations (unless a very high order is used). Numerical experiments using 2D-HYDROTUR showed that the "plunging jet" begins to develop for the depth ratio 1.20 . This explains the value 1.19 found by Abbott & Rodenhuis (1972).

6.4.3 Turbulent hydraulic jump.

In this section, numerical calculations of two moving turbulent jumps with depth ratios 2.17 and 3.52 are described. The two depth ratios were obtained by combining the acceleration of gravity with the input stream velocity to impose the appropriate Froude number. The purpose of these simulations was to test the full capabilities of 2D-HYDROTUR in fully developed free-surface turbulent flows. The first of these jumps is representative of broken waves in the inner region (for which $H = \mathcal{O}(D)$), whereas the second (stronger) jump was included in the present work to allow a more realistic comparison with the experimental results of Resh & Leutheusser (1972) (which provide a useful orientation for interpreting the results of both simulations).

The first turbulent hydraulic jump was generated using the following data. The acceleration of gravity was chosen $-0.25\ ms^{-2}$, the input stream velocity was $-0.5\ ms^{-1}$ and the uniform stream depth was $1.0\ m$. This gives $\mathbf{F}=1.854$, $h_1/h_2=2.17$ and $C=0.427\ ms^{-1}$. The domain was discretized using a uniform mesh of 180×45 cells. The molecular viscosity was $1.3\times10^{-6}\ m^2s^{-1}$ (the viscosity of salt water at $10^\circ C$). Standard values were assigned to the constants in the turbulence model. The boundary conditions for the dependent variables were prescribed using the law-of-the-wall at all boundaries except the right, where continuative (open) boundary conditions were prescribed. This allowed a continuous input of fluid with low-intensity turbulence.

Figures 6.25–6.27 illustrate the velocity field and surface profile, the distribution of ν_t/ν (turbulence Reynolds number), and the distribution of K, respectively, for the time $t=35$ s, when the computed jump became almost stationary in a frame of reference moving to the right with velocity C. The velocity plot shows an extensive shear layer originated by the folding of the free surface, starting at the toe of the jump and ending at a distance of (approximately) $4h_1$ downstream. Within this shear layer there was a primary recirculation region in which the turbulence intensity was maximum. Two secondary recirculation zones with non-negligible vertical velocity were found at $2h_1$ and $4h_1$ downstream of the jump's front. The occurrence of these "roll vortices" is well documented in the literature of surf-zone hydrodynamics. Several small void regions were observed inside the body of the jump, caused by the violence of the initial breaking. In reality, strong air concentrations would be found instead. At $t=35$ s the computed height of fluid at the wall was 2.20, with an error of 1.4% to the theoretical value 2.17 .

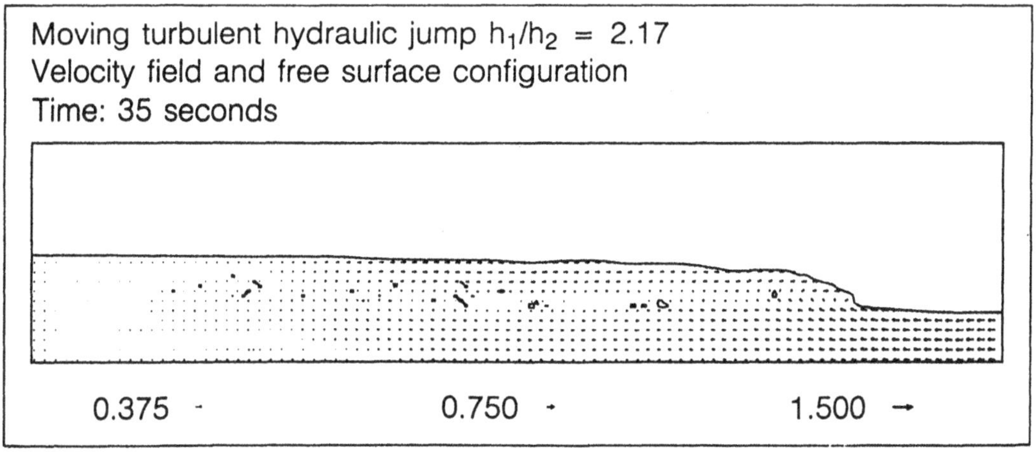

Figure 6.25 : *Computed velocity field and free-surface configuration for a moving hydraulic jump with depth ratio* 2.17 .

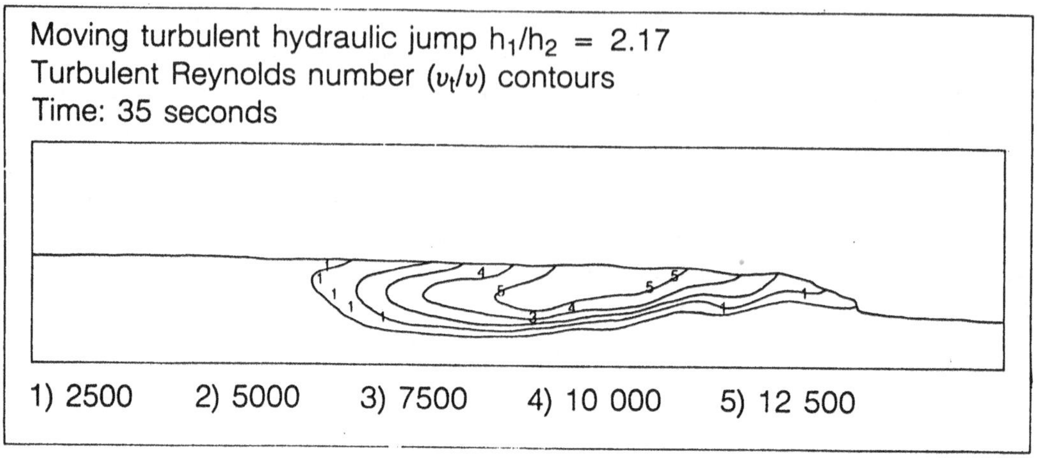

Figure 6.26 : *Computed distribution of dimensionless eddy-viscosity (turbulence Reynolds number)* ν_t/ν *for a moving hydraulic jump with depth ratio* 2.17 .

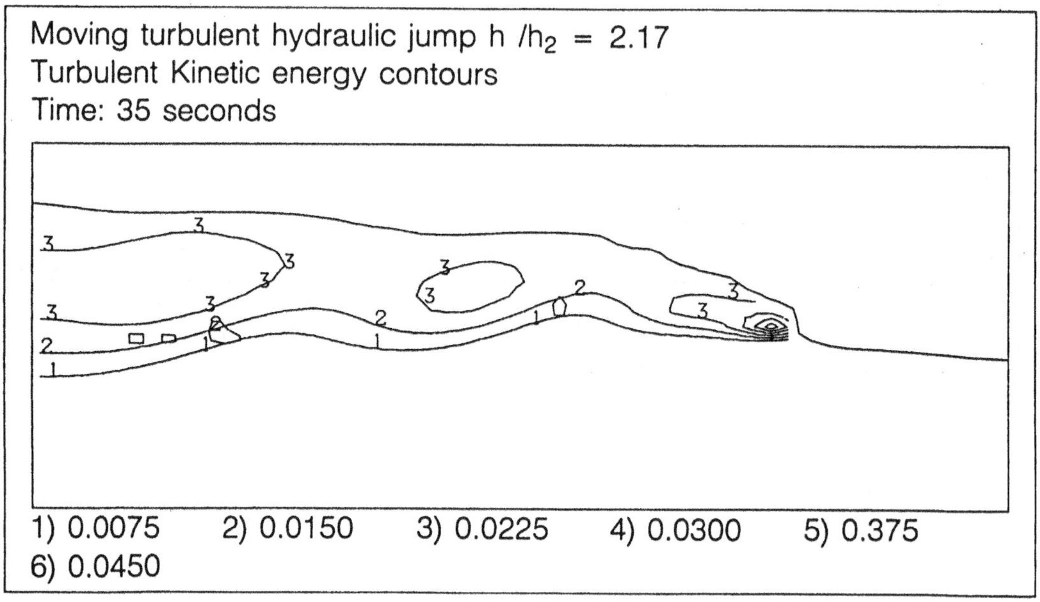

Figure 6.27 : *Computed distribution of turbulent kinetic energy (K) for a moving hydraulic jump with depth ratio* 2.17 .

The relevant turbulence dynamics was confined to the surface shear layer, the bottom-generated turbulence being irrelevant (as expected). In this surface layer, values of eddy-viscosity ranging from 3.0×10^{-3} to 1.5×10^{-2} m^2s^{-1} were found, and the turbulence Reynolds number showed strong gradients. The maximum of eddy-viscosity was found downstream of the main production region, showing that ϵ decays relatively faster than K. There was strong production at the toe of the front, where K reached an absolute maximum of about 10% of the kinetic energy of the incident stream. The turbulence generated close to the jump front spread downward and downstream, justifying the qualitative model of Madsen & Svendsen. Most of the turbulence was dissipated above the height of the incident stream, *i.e.* at the same level where it was generated.

Figures 6.28–6.30 illustrate the computed distributions of the non-dimensional quantities $\sqrt{(\overline{u^2})}/|u_2-C|$, $\sqrt{(\overline{v^2})}/|u_2-C|$ and $\nu_t(\partial u/\partial y)/|u_2-C|^2$, which are functions of the supercritical Froude number. The standard K-ϵ model cannot predict the relative strength of the normal stresses. In this simulation, relative strengths appropriate for turbulent wakes were assumed (according to Svendsen's (1987) suggestion).

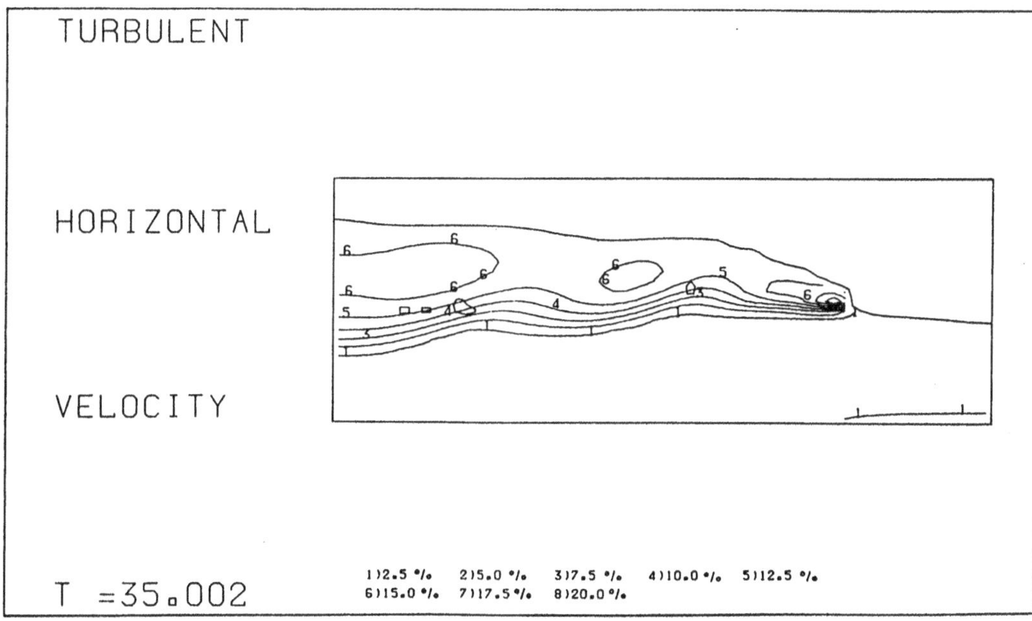

Figure 6.28 : *Computed distribution of dimensionless horizontal velocity fluctuation for a moving hydraulic jump with depth ratio* 2.17 .

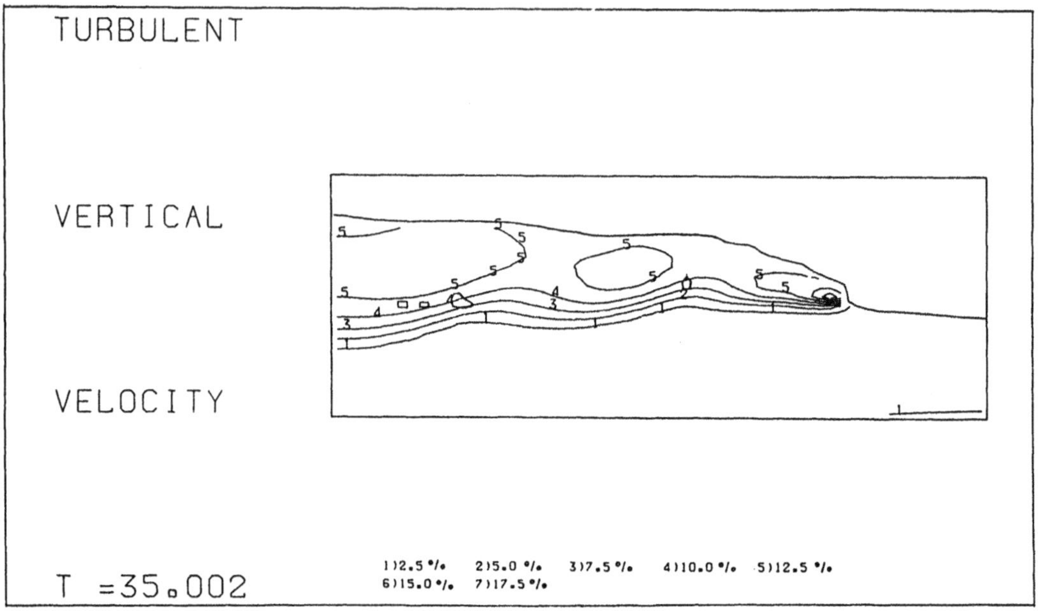

Figure 6.29 : ***Computed distribution of dimensionless vertical velocity for a moving hydraulic jump with depth ratio 2.17 .***

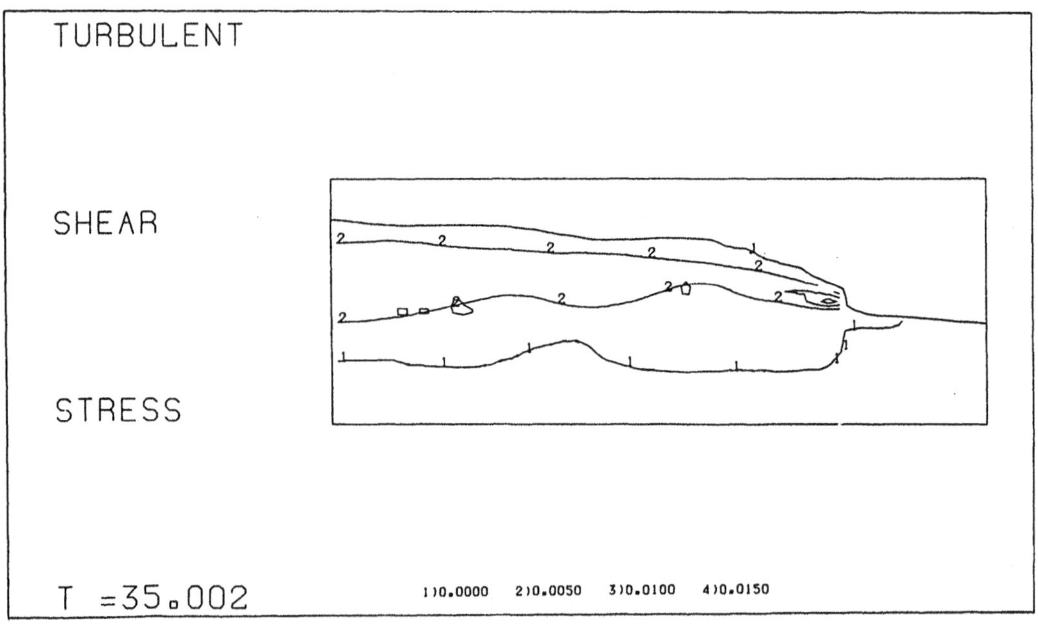

Figure 6.30 : ***Computed distribution of dimensionless shear stress for a moving hydraulic jump with depth ratio 2.17 .***

Although the distributions of the three dimensionless quantities near the center of the shear layer were within a few percent of the values found experimentally (see figure 3.11, chapter 3) by Resh & Leutheusser (1972), the downward spread and downstream decay of the turbulence field were smaller than expected. To provide a more correct comparison with the above mentioned experimental study, a second, stronger jump with $\mathbf{F}=2.83$ and depth ratio 3.52 was simulated. This supercritical Froude number was obtained using a more realistic combination of physical parameters. The acceleration due to gravity was chosen $g=9.81$ ms^{-2} downwards, and the input stream velocity was $u_2=-2.0$ ms^{-1}. The computational domain was chosen 2.5 m long and 0.5 m high. The domain was discretized using a regular mesh of 200×40 cells.

Figures 6.31–6.32 show a comparison between the computed and experimental distributions of the dimensionless turbulent velocity fluctuations and shear stress at distances $2.0h_1$ and $3.0h_1$ downstream of the front of the jump. The agreement was fair but there were significant differences. First, near the bottom the computed turbulence intensity was smaller than the experimental values, because bottom-generated turbulence is stronger in steady jumps than in moving jumps. Secondly, the experiments show that the relative strength of the turbulent velocity fluctuations is not constant inside the jump. This aspect of the turbulence dynamics can only be described by using a more sophisticated turbulence model. Finally, the computed downstream decay of the turbulence variables is less pronounced than was found in the experiments.

The computed depth ratio for this jump was within 1% of the theoretical value , showing again good overall momentum conservation. Therefore, it seems that the incorporation of the turbulence model in the NS solver does not affect the accuracy of momentum conservation in free-surface flows. This is an important characteristic of the present model.

The study of the turbulent jumps provided a good verification of the predictive capabilities of the present model for strongly turbulent free surface flows.

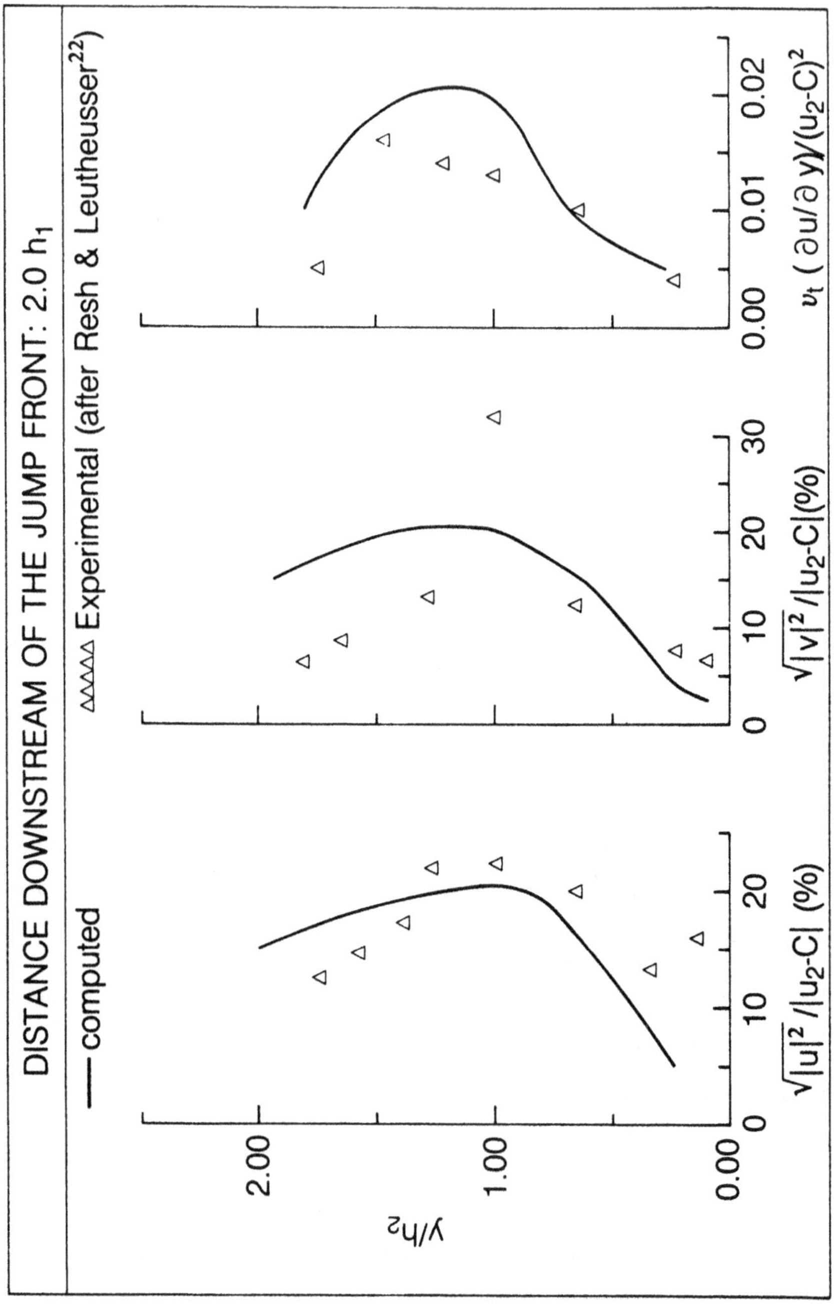

Figure 6.31 : ***Computed vs. experimental distributions of the dimensionless normal and shear stresses for a hydraulic jump with*** **F**=2.83 , ***at a distance*** $2h_1$ ***downstream of the jump front. Experimental data after Resh & Leutheusser*** (1972).

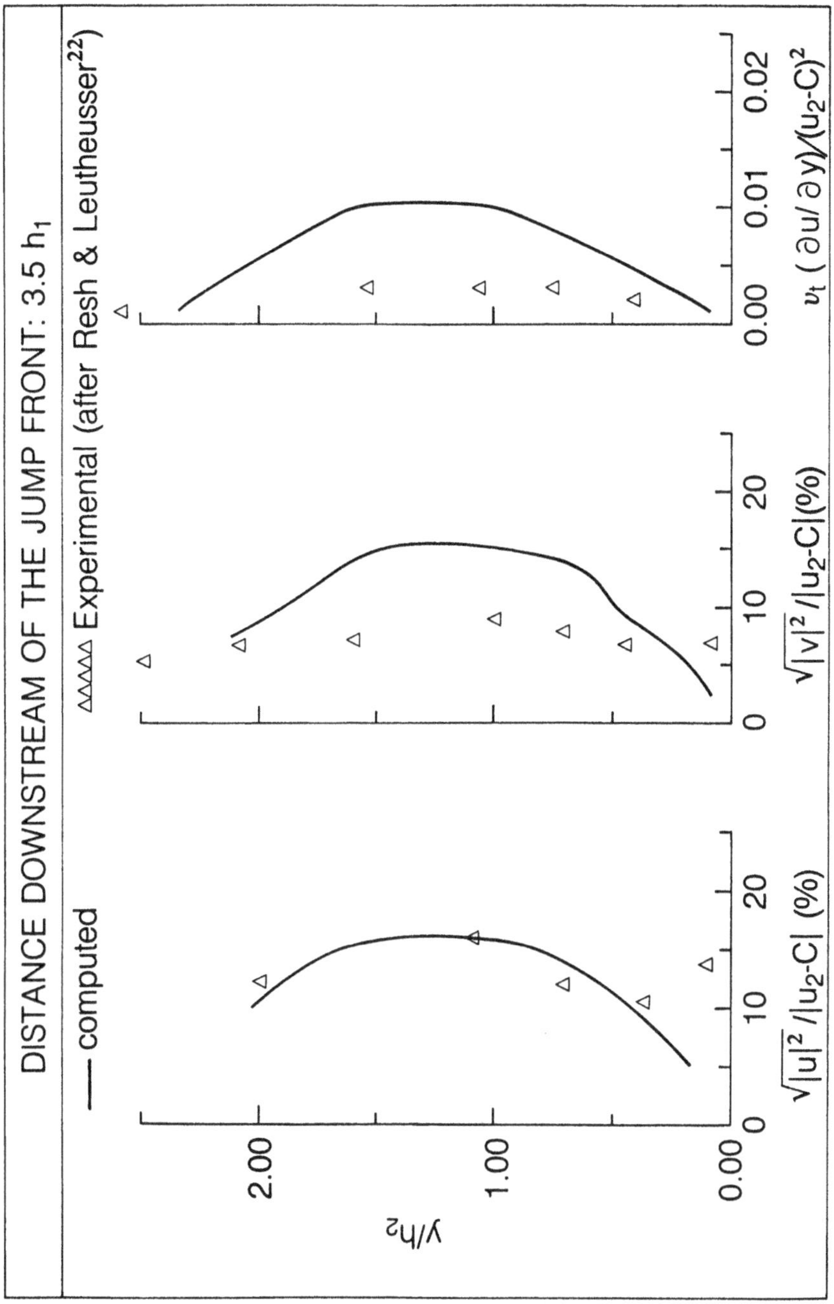

Figure 6.32 : *Computed vs. experimental distributions of the dimensionless normal and shear stresses for a hydraulic jump with* $\mathbb{F}$=2.83 , *at a distance* $3.5h_1$ *downstream of the jump front. Experimental data after Resh & Leutheusser* (1972).

6.5 Breaking of a solitary wave over a slope.

As a first attempt to describe the whole sequence of wave transformation processes across a surf-zone, a numerical simulation of the breaking of a solitary wave over a slope was performed. The results of this experiment will be described in this section, with emphasis on the fully turbulent stage. From a physical viewpoint, the main difficulty of this problem is the sudden transition from quasi-irrotational, nearly laminar motion to highly turbulent, rotational flow. This leads to the existence of regions in the fluid domain where the gradients of the dependent variables are very strong, therefore introducing problems of numerical stability.

The physical domain was chosen 12 m long and 0.5 m high, and consisted of three sections (see figure 6.33). The first section, from $x=0$ to $x=4$ m, had a horizontal bottom and a still water depth of 35.5 cm. The second section, from $x=4$ to $x=8.3$ m, had an average bottom slope of 1/20 which was defined by blocking out mesh cells. The third section, from $x=8.3$ to $x=12$ m, had a horizontal bottom and a still water depth of 14 cm. This setup reproduced approximately the physical conditions of one of Nadaoka's experiments.

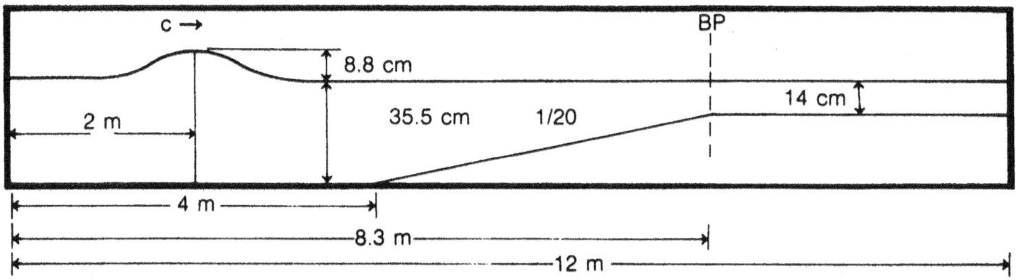

Physical parameters	Geometrical parameters		Discretization parameters
$g = 9.81$	length of domain	12m	$I = 500$ cells
$\upsilon = 1.3\times10^{-6}$	Height of domain	0.5m	$J = 50$ cells
$K_i = 2.62\times10^{-3}$	Dimensionless		$\Delta y = 1.0$ cm
$\epsilon_i = 1.07\times10^{-3}$	Wave height	0.25	$\Delta x = 2.4$ cm
$[\upsilon_t/\upsilon \simeq 45]$			$\Delta t = 2.5\times10^{-4}$ s

Figure 6.33 : *Setup for the solitary wave problem.*

The initial and boundary conditions were the following. A solitary wave with dimensionless height 0.25 and crest at $x=2$ m was generated by specifying both velocity components, pressure and F-function values according to the third order solitary wave theory. The initial conditions for K and ϵ were $K=2.62\times10^{-2}$ m^2s^{-2} and $\epsilon=1.07\times10^{-3}$ m^2s^{-3}

($\nu_t/\nu \simeq 45$). On the bottom, wall boundary conditions were imposed. The remaining boundaries were treated as open. The fluid molecular viscosity was 1.3×10^{-6} m^2s^{-1} and the acceleration of gravity was 9.81 ms^{-1} downwards. The domain was discretized using a regular mesh of 500×50 cells. The time step was $\Delta t = 2.5\times10^{-4}$ s .

After 1 second of propagation (not shown) the wave reached the slope. The wave profile was nearly symmetrical and the velocity (relative to the crest) was almost identical to the initial condition. The turbulent kinetic energy, the eddy viscosity, and the other turbulence-related variables were significant only in two thin regions—boundary layers—one near the bottom and the other near the wave crest. In these two boundary layers, the turbulent kinetic energy was about 1% of the kinetic energy of the mean flow, and the turbulence Reynolds number ν_t/ν was $\simeq 50$. Vorticity was strongly concentrated in the two boundary layers, as expected. The turbulent shear stress was residual and spatially uncorrelated, except in the two boundary layers, where $\nu_t(\partial U/\partial y)/U^2 \simeq 2\times10^{-3}$.

After 3 seconds of propagation (not shown) the wave reached the last section of the domain and was at the onset of breaking. Its profile was then highly deformed, with a very steep front face. This configuration was accompanied by the formation of a plunging jet, with fluid velocities higher than $(gD)^{1/2}$. Turbulent kinetic energy increased 150% with respect to the pre-breaking stage, but remained 0.6% of the local kinetic energy of the mean flow in the region where the plunging jet was forming. The turbulence Reynolds number was then $\nu_t/\nu \simeq 125$ in the two boundary layers, but remained close to the initial condition in the bulk of the fluid.

Figure 6.34 shows the mean velocity and the turbulent kinetic energy for the broken wave after 4.5 seconds of propagation. A strong shear layer was formed near the surface, as a consequence of the collapse of the plunging jet. The turbulence intensity in this shear layer was high, about one order of magnitude larger than at the onset of breaking. However, the downward spread of turbulence was small.

Figure 6.35 shows the distributions of the turbulence Reynolds number ν_t/ν and the turbulent shear stress for the broken wave at the same instant. The spatial distribution of these variables was similar to that of the turbulent kinetic energy, and their maxima were two orders of magnitude larger than before breaking. The model represented the essential features of the transition from pre-breaking to broken wave motion. However, the transition from low to high intensity turbulence after breaking was less abrupt than expected. The decay of the dissipation rate ϵ produced too high values of the eddy-viscosity before breaking. It was found that the distributions of the turbulence variables in fully developed stage were more concentrated and had higher maxima than those found by Nadaoka (1986) in his experiments with periodic waves in otherwise similar physical conditions.

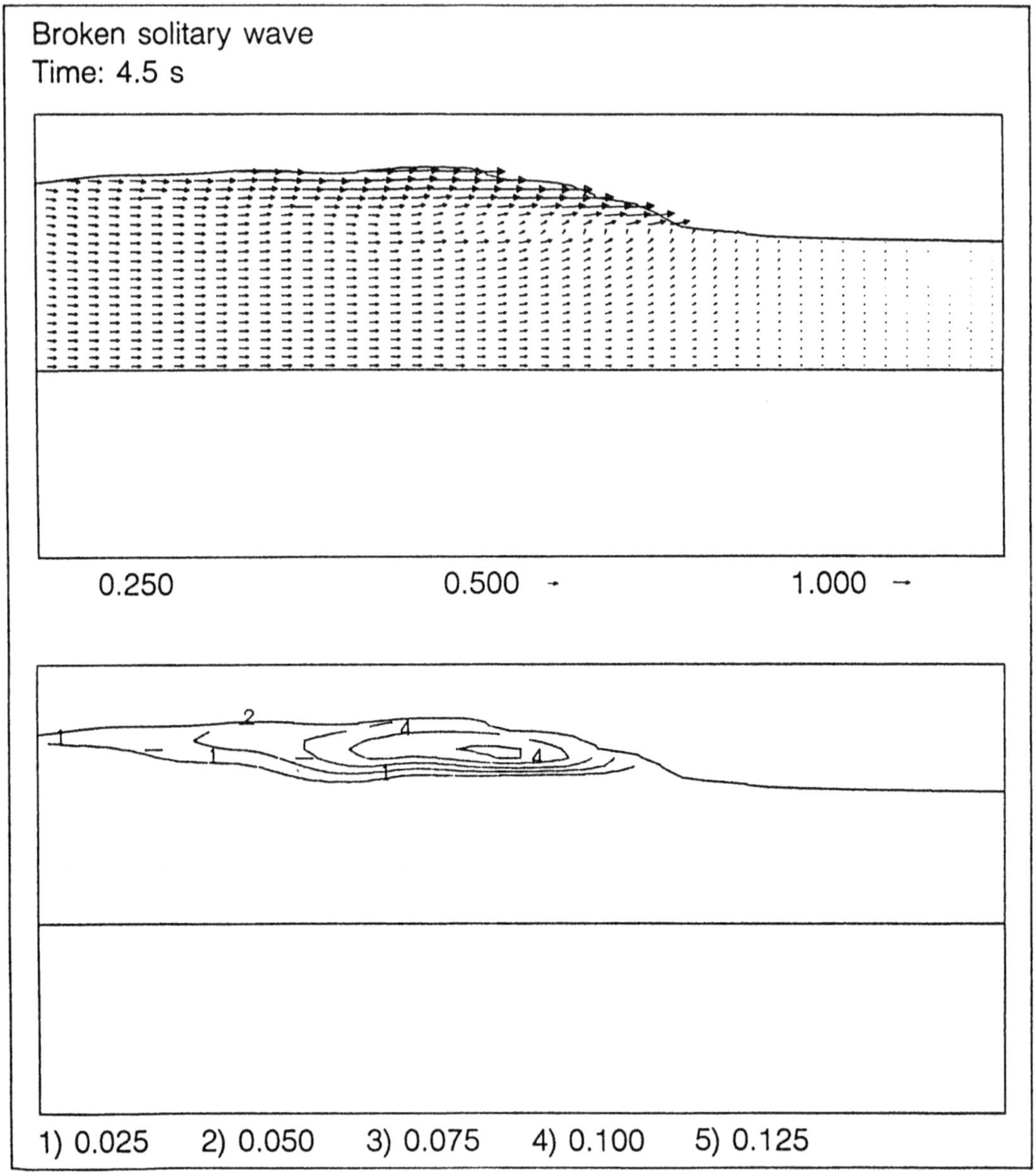

Figure 6.34 : *Velocity field (top) and turbulent kinetic energy (bottom) for the broken solitary wave after 4.5 seconds of propagation.*

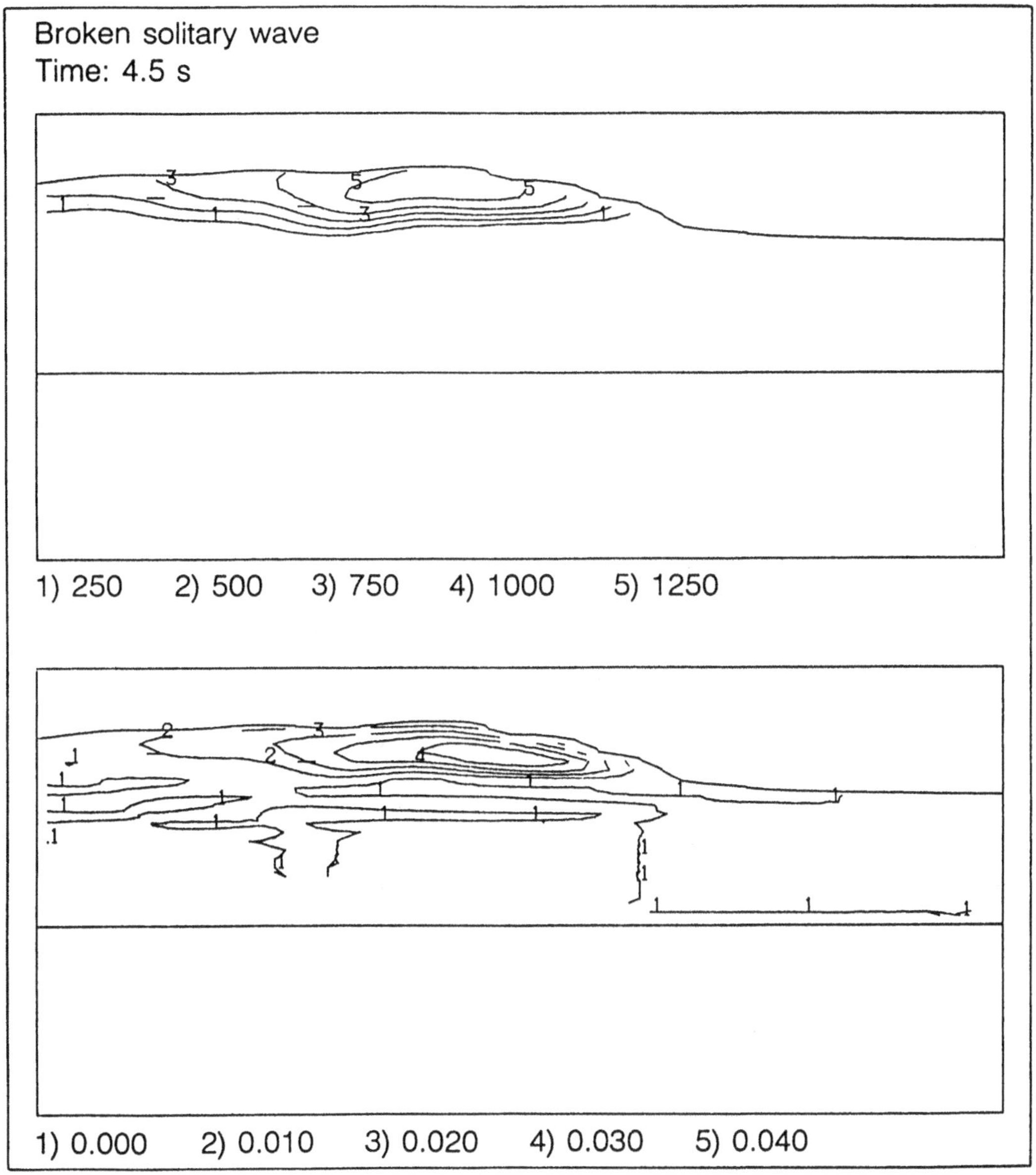

Figure 6.35 : *Turbulence Reynolds number* ν_t/ν *(top) and turbulent shear stress (bottom) for the broken solitary wave after 4.5 seconds of propagation.*

The computed distributions of mean vorticity, mean rate of deformation and turbulent shear stress were also more concentrated and had higher maxima than the corresponding distributions found by Nadaoka. There are several explanations for this discrepancy, apart from the deficiencies of the K-ϵ model. First, it should be noted that the experimental investigations also have limitations. For instance, many of the results reported by Nadaoka (1986) were obtained by using a moving average method for separating the turbulent component from the periodic one. This method underestimates the energy of the large eddies (Nadaoka, 1986). In contrast, in the present computation large eddies were not observed, and thus their energy is represented as turbulent kinetic energy by the K-ϵ model. An indirect demonstration of this is given in figure 6.36 . This figure illustrates the breaking of a solitary wave over a slope of 1 on 20. An effective (physical plus numerical) viscosity of 10^{-4} m^2s^{-1} (one to two orders of magnitude lower than the typical eddy-viscosity for such a problem) was imposed, and K-ϵ model was not used. The collapse of the overturning motion into large eddies was then well captured.

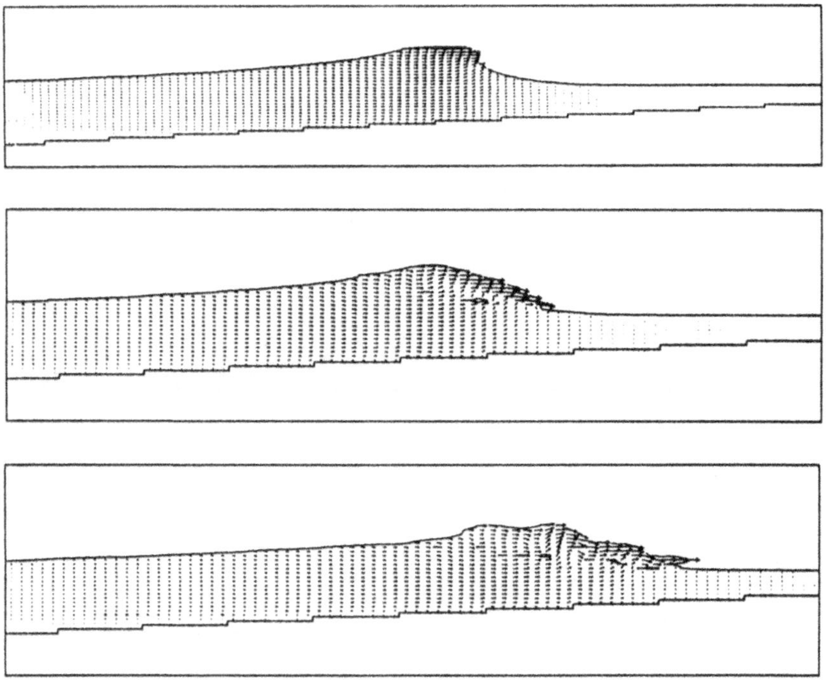

Figure 6.36 : *Collapse of a overturning solitary wave into large eddying motion.*

Another reason why the calculations showed higher maxima and more concentrated distributions, is that the maxima were found in regions of the flow where aeration rules out precise experimental measurements. Another reason for the observed discrepancies is that the turbulence generated by the breaking of periodic waves is not identical to that generated by the breaking of a single wave.

6.6 Breaking of a train of periodic waves over a slope.

The last calculation reported in this work, is a numerical simulation of the breaking of a train of periodic waves over a slope. The main purpose of this numerical experiment was to compare the dynamics of broken periodic waves, with the dynamics of broken solitary waves. Periodic waves give a better representation of reality than solitary waves, because some relevant phenomena, like the distribution of wave-averaged properties, can only be studied in terms of periodic or "quasi-periodic" waves.

In this experiment, the physical conditions were specified as follows. The computational domain was chosen 9 m long and 0.40 m high, with a still water level of 0.30 m . Periodic waves were generated at the left boundary by means of special boundary conditions for u, v and F using second-order Stokes theory. This wave train had the following parameters: height $H=$ 6 cm and period $T=1.22$ s . Starting at $x=2$ m from the left boundary a slope of 1/20 was defined approximately by blocking out mesh cells. The waves broke over this slope, as a consequence of nonlinear steepening. The acceleration due to gravity was $g=9.81$ ms^{-2} and the molecular viscosity was chosen $\nu=1.3\times10^{-6}$ m^2s^{-1} . This setup is similar to that of Mizuguchi's (1986) experiments. The initial conditions for the turbulence variables were $K=2.253\times10^{-5}$ m^2s^{-2} and $\epsilon=8.574\times10^{-6}$ m^2s^{-3} everywhere, corresponding to very low initial turbulence intensity. The computational domain was discretized using a regular mesh of 300×40 cells, with $\Delta x=3$ cm and $\Delta y=1$ cm.

Figure 6.37 shows the mean velocity field and the free surface configuration after several waves have been introduced in the computational domain. This figure also illustrates the typical sequence of transformations suffered by wind waves across a surf-zone. In the initial stages of shoaling, the waves remain nearly symmetrical (waves nrs. 1 and 2). Because of nonlinear steepening, the front face of the waves becomes steep; a plunging jet where the particle velocity exceeds $(gD)^{1/2}$ is formed near the wave crest (wave nr. 3). After breaking, the waves propagate much like moving hydraulic jumps (wave nr. 4). Finally, a "run-up" region is observed near the shoreline, where the mean water level is slightly higher than offshore to equilibrate the loss of wave momentum flux (or radiation stress) across the surf-zone (wave nr. 5). Figure 6.38 shows very clearly the fundamental difference between the velocity field of pre-breaking and broken waves. Before breaking the velocity field remains nearly irrotational and symmetrical about the wave crest. After breaking there is a strong shear layer above the trough level and vorticity is high.

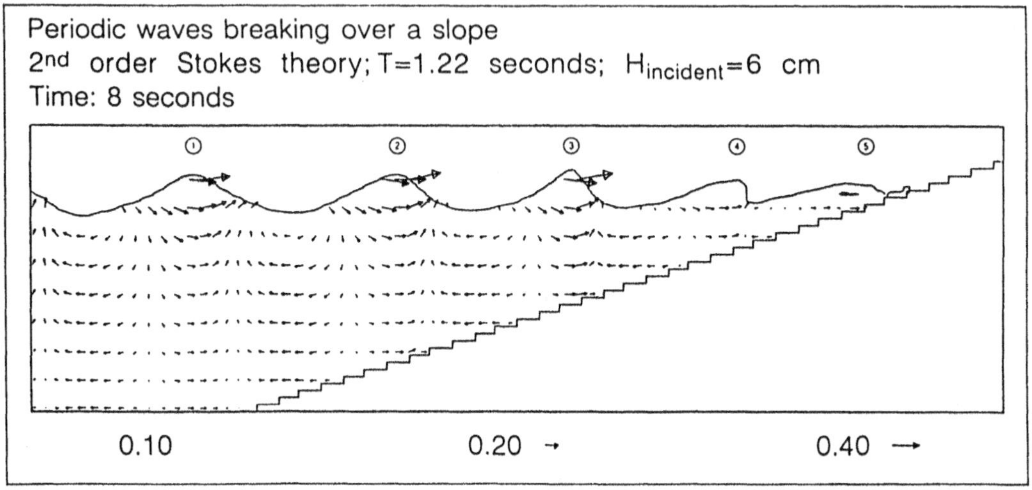

Figure 6.37 : *Mean velocity and surface configuration for a train of periodic waves.*

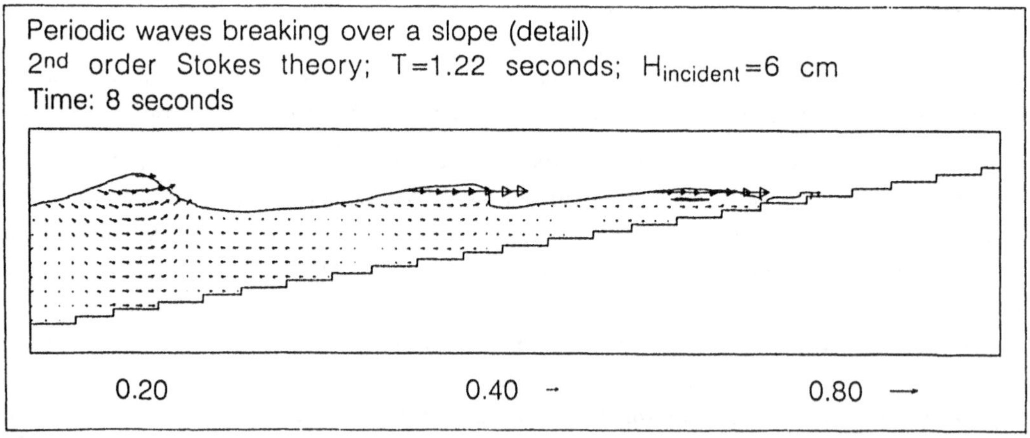

Figure 6.38 : *Detail of the velocity field in the surf-zone between x=4.5 and x=9 m, at t=8 s .*

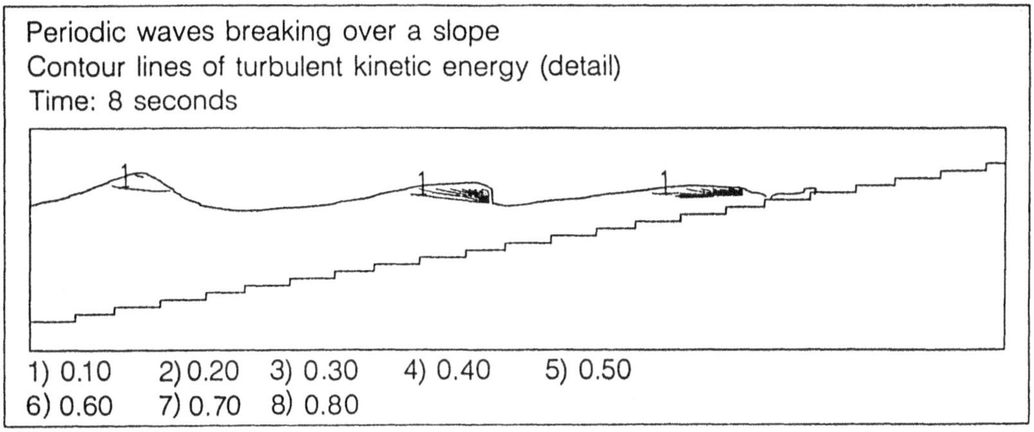

Figure 6.39 : *Contour lines of turbulent kinetic energy in the surf-zone between x=4.5 and x=9 m , at t=8 s .*

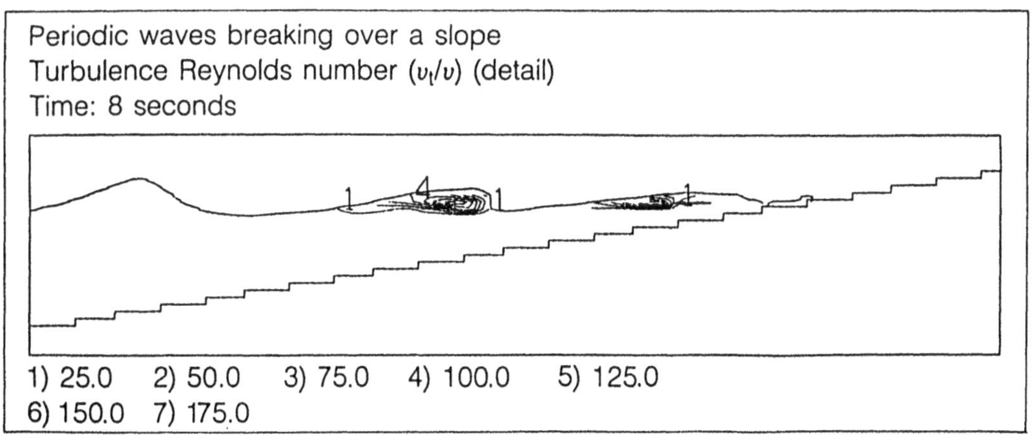

Figure 6.40 : *Contour lines of turbulence Reynolds number ν_t/ν in the surf-zone between x=4.5 and x=9 m , at t=8 s .*

Figures 6.39 and 6.40 show the contours of turbulent kinetic energy and turbulence Reynolds number ν_t/ν in the surf-zone. As expected, turbulence was much more intense after breaking. The regions with high turbulence intensity were confined to the region above the trough level. As in the previous simulation, the computed distributions of the turbulence variables were more concentrated and had higher maxima than was found in physical experiments (*e.g.* Mizuguchi, 1986). In this respect, the differences between the breaking of solitary and periodic waves were smaller than expected.The computed results suggest that surf-zone waves behave somewhat like a train of independent solitary waves, with a small return flow under their troughs. Most of the turbulence is dissipated where it is created. The downward spread of turbulence intensity by advective and diffusive transport seems to be relatively slow.

Chapter 7

CONCLUSIONS
FUTURE RESEARCH AND DEVELOPMENT

7.1 Summary and conclusions.

A new methodology for a unified treatment of wave transformation before, during and after breaking was developed. This methodology was elaborated as follows. First, the essential aspects of the hydrodynamics of incompressible flows were reviewed, with special reference to turbulence dynamics and free-surfaces. Then, the physical aspects of wave transformation processes were analyzed, the limitations of the available techniques were discussed, and the structure of the present model was outlined. Finally, the numerical model was constructed and applied to several wave transformation problems, demonstrating its predictive capabilities for a unified description of the flow before, during and after breaking.

The most important results and conclusions obtained in each chapter are now summarized.

Chapter 2

A basic review of incompressible flow hydrodynamics was presented, and the turbulent transport processes were analyzed by means of transport equations. It was concluded that the K-ϵ model is a good compromise between cost and effectiveness. Several types of boundary conditions for the dependent variables were discussed, as well as their limitations.

Chapter 3

The physical processes of shallow water wave transformation were analyzed in detail, and the elements of the present methodology were outlined. These consist of the numerical integration of the incompressible Reynolds equations (momentum and continuity equations for the turbulence-averaged flow) with Boussinesq closure, the modeled K and ϵ equations, and a technique for treating time-dependent multiple free-boundaries. The limitations of the proposed methodology (particularly of the incompressibility assumption) were also discussed. The algorithm so defined was seen to be an extension of MAC-type methods for incompressible viscous fluid flow involving free-surfaces.

Chapter 4

MAC-type algorithms were reviewed in a comparative manner, with emphasis on the efficiency of pressure solution and on the technique for defining and updating the free-surface. It was found that the pressure solution is best treated using the pressure-velocity relaxation technique for the pressure Poisson equation. It was also found that the most efficient technique for treating multiple free-surfaces of arbitrary configuration is the VOF (Volume-of-Fluid) technique.

Chapter 5

The discrete equations of the model were presented. A hybrid upwind/modified FTCS scheme was used for the momentum equations.

Several approximations for the K-ϵ equations were considered, with the aim of avoiding excessive numerical diffusion and the stiffness of the source/sink terms, while keeping the structure of the basic algorithm as simple as possible (to improve the possibilities for vector and parallel optimization).

Simple boundary conditions for K and ϵ at a free-surface of arbitrary configuration were derived and discussed, and an efficient method for computing wall boundary conditions was introduced.

A heuristic (objective analysis) method for specifying initial conditions for the K-ϵ equations was introduced, which is especially useful when experimental information is unavailable and theoretical solutions are unsuitable for the problem at hand.

A heuristic stability analysis of the K-ϵ equations was performed. This analysis showed the formal similarity of the modeled K and ϵ equations, and the relationship between the numerical problem of stability and the physical problem of breakdown of the K-ϵ model.

The numerical model was tested in laminar and turbulent flows of theoretical or engineering interest, confined and with free-surfaces. Analytical and numerical solutions were available for comparison, and good performance was found in all test problems.

Chapter 6

It was shown that the present model describes accurately pre-breaking phenomena, transition from dispersive to dissipative regime, and fully developed turbulent flow. Therefore, problems that required different approaches and solution techniques can now be investigated using a single methodology.

It was found that the order of approximation of the wave theory used for generating initial and boundary conditions has a greater influence in the quality of the solutions for highly nonlinear waves than truncation errors.

The nature of the transition from undular (dispersive) to turbulent (dissipative) regime was studied using the hydraulic jump as a basic model. The critical depth ratio for breaking was calculated, and it was found that transitional hydraulic jumps are similar to other waves at the onset of breaking.

Two turbulent hydraulic jumps were simulated, and the computed distribution of the relevant turbulence field variables were in good agreement with experimental results of Resh & Leutheusser (1972). This provided a verification of the predictive capabilities of the present model for strongly turbulent free-surface flows.

Two numerical experiments on the transition from pre-breaking to broken wave motion were performed, one using a solitary wave, the other using a train of periodic waves. The main features of the solution were analyzed by providing computed plots of velocity, turbulent kinetic energy, turbulence Reynolds number, turbulent velocity fluctuations, and turbulent shear stress. The qualitative and quantitative features of both numerical solutions were almost identical:

• First, it was found that the model represents transition from nearly irrotational to fully turbulent flow, but overestimates the turbulence level in the first stages of the transformation. This effect is more pronounced for ϵ than for K or the mean flow variables (vorticity and rate of deformation).

• Second, in the fully developed flow stage, the model yielded more concentrated distributions of turbulence intensity, turbulent shear stress, mean vorticity and mean rate of strain near the breaking front than were found in physical experiments. This discrepancy is due to limitations of the present K-ϵ model. Nevertheless, it should be noted that the experimental investigations also have important limitations. Thus, a possible future application of the present approach is to serve as a complement to experimental studies, somewhat as a "numerical laboratory".

The computer program (2D-HYDROTUR) is highly optimized and reliable. Optimization features include global vector optimization and vector/parallel optimization, thus providing an efficient use of the machine on which the program is installed. Program reliability has been proven by running it in several machines (IBM-PC, HP 9000/350, VAX-8600, ALLIANT FX-40 vector/parallel processor, and CONVEX C200 parallel computer). In vector computers, the performance is strongly dependent on the structure of the program. With the vector-optimized version of the code, a time factor of 2.2 with respect to the scalar version was attained in an ALLIANT FX-40 vector/parallel computer, for which the optimum time factor is $\simeq 5$ (unless the structure of the program is very simple).

7.2 Future research and development.

A remarkable quality of the present model is the exceptional promise it holds for future refinements. Future research and development of the present work should go in two directions: improvement of the algorithm (especially the turbulence model), and application to more sophisticated problems.

Improvements of the algorithm should first aim at the elimination of the problems that have been found in the applications, and then at the improvement of the predictive capabilities of the model. Some of these improvements are :

• Introduction of arbitrary cell boundaries, to eliminate artificial roughness problems. From the numerical viewpoint, artificial roughness is the most important spurious effect found so far, which disrupts the structure of the flow near obstacles that are not aligned with the mesh. This will be eliminated in a near future, by using a refinement of the pressure-velocity iteration step which allows arbitrary cell boundaries while maintaining the same type of finite difference mesh (Viecelli, 1969, 1971). Also, the use of more efficient discretizations for the transport equations is being considered at the time of writing.

• Introduction of density variation and limited compressibility effects, to take into account air-water mixing. Inclusion of these two effects in the present model would require an additional transport equation for the air concentration, and a partial restructuration of the algorithm. Work is presently under way to incorporate these refinements.

• Extensive calibration of the constants in the K-ϵ model, and introduction of new closure schemes that replace these constants by functions of the local state of turbulence, to obtain a more realistic transition from low to high intensity turbulence.

• Refinement of the K-ϵ model to allow for anisotropic diffusion and streamline curvature.

• Refinement of the free-surface turbulence boundary conditions by taking into account the actual configuration of the boundary.

• Incorporation of a K-ϵ-A or full second order turbulence closure model to the existing algorithm.

• Extension of the algorithm to three spatial dimensions.

The incorporation of these improvements must preserve the efficient optimization of the program for scalar and vector processing, otherwise the practical interest of the present methodology may be lost. Another possible improvement is the vectorization of the portions of the program that are presently operating in scalar mode (this would have little—if any—effect on scalar computers).

The capabilities of the present model (and future refined versions) should be tested in practice by applying it to more sophisticated problems. Some of these problems are :

• Study of the properties of collisions between highly nonlinear waves, and comparison with physical or numerical experiments.

• Study of wave overturning, and comparison with results obtained with time-dependent potential flow models.

• Study of turbulence distribution in broken waves, in conjunction with physical experiments.

• Numerical calculation of the distributions of global (integral) wave properties, and comparison with state-of-the-art theories. This application is particularly important from the viewpoint of coastal engineering, since most operative models for predicting surf-zone currents are based on wave-averaged equations.

• Study of wave interaction with coastal and offshore structures.

• Ship hydrodynamics, if a 3D version of the code becomes operative.

REFERENCES

Abbott, M. B. (1979) . Computational Hydraulics : Elements of the theory of free-surface flows. Pitman.

Abbott, M. B. and Rodenhuis, C. S. (1972) . A numerical simulation of the undular hydraulic jump. Jour. Hyd. Res. , vol. 10 , pp 239-257.

Abbott, M. B. and Rodenhuis, C. S. (1974) . Discussion. Jour. Hyd. Res. , vol. 12 , pp 141-152 .

Amsden, A. A. and Harlow, F. H. (1968) . Transport of turbulence in numerical fluid dynamics. Jour. Comp. Phy. , vol. 3 , pp 94-110.

Amsden, A. A. and Harlow, F. H. (1970) . The SMAC method : A numerical technique for calculating incompressible fluid flows. Los Alamos Scientific Laboratory, report LA-4370.

Argyris, J. and Haase, M. (1987) . An engineer´s guide to soliton phenomena: application of the finite element method. Comp. Meth. Appl. Mech. Eng. , vol. 61 , pp 71-122.

Baker, G. R. , Meiron, D. I. and Orszag, S. A. (1982) . Generalized vortex methods for free-surface flow problems. Jour. Fluid Mech. , vol. 123 , pp 477-501.

Batchelor, G. K. (1953) . The theory of homogeneous turbulence. Cambridge University Press.

Batchelor, G. K. (1967) . An Introduction to Fluid Dynamics. Cambridge University Press.

Battjes, J. A. (1974) . Surf similarity. Proc. Int. Conf. Coastal Eng. 14th, pp 466-479.

Battjes, J. A. (1988) . Surf-zone dynamics. Annual review of fluid mechanics, Palo Alto, California, pp 257-293.

Bélanger, J. M. (1828) . Essai sur la Solution Numérique de Quelques Problémes Relatives au Movement Permanent des Eaux Courantes. Paris.

Benjamin, T. B. and Feir, J. E. (1967) . The disintegration of wave trains in deep water. I. Theory . Jour. Fluid Mech. , vol. 27 , pp 417-430.

Benjamin, T. B. and Lighthill, M. J. (1954) . On cnoidal waves and bores. Proc. Roy. Soc. Lond. A 224, pp 448-460.

Boris, J. P. and Book, D. L. (1976) . Solution of the continuity equations by the method of flux corrected transport. Methods in Comp. Phy. , vol. 16 , Academic Press.

Bradshaw, P. (1978) . Chapter 1 of *Turbulence*, ed. P. Bradshaw . Springer Verlag.

Bulgarelli, U. , Casulli, V. and Greenspan, D. (1984) . Pressure methods for the numerical solution of free surface fluid flows. Pineridge Press, U.K. .

Byatt-Smith, J. G. B. (1971) . An exact integral equation for unsteady surface waves and a comment on the Boussinesq equation. Jour. Fluid Mech. , vol. 49 , pp 625-633.

Carrier, G. F. and Greenspan, H. P. (1958) . Water waves of finite amplitude on a sloping beach. Jour. Fluid Mech. , vol. 4 , pp 97-107.

Casulli, V. and Greenspan, D. (1988) . Numerical analysis for applied mathematics, science and engineering. Addison Wesley Publishing Company.

Celik, I. , Rodi, W. and Stamou, A. T. (1987) . Prediction of hydrodynamic characteristics of rectangular settling tanks. In *Turbulence measurements and flow modelling*, eds. C. J. Chen et al. , pp 641-651 . Springer Verlag.

Cokelet, E. D. (1977) . Steep gravity waves in water of arbitrary uniform depth. Philos. Trans. Royal Soc. London A 286, pp 183-230.

Cooker, M. J. , Peregrine, D. H. , Vidal, C. and Dold, J. W. (1989) . The interaction between a solitary wave and a submerged semi-circular cylinder. Universidad de Cantabria, 39005 Santander, Spain. Report No. AM-89-04, June 1989.

Cousteix, J. , Desopper, A. , and Houdeville, R. (1977) . Structure and development of a turbulent boundary-layer in an oscillatory external flow . In *Turbulent Shear Flows I* , eds. F. Durst et al. , pp 154-171 . Springer Verlag.

Chan, K. C. and Street, R. L. (1970) . A computer study of finite-amplitude water waves. Jour. Comp. Phy. , vol. 6 , pp 68-94 .

Chen, C. J. and Chang, S. M. (1987) . Prediction of turbulent flows in rectangular cavity with K-ϵ-A and K-ϵ-E models. In *Turbulence measurements and flow modelling* , eds . C. J. Chen et al. , pp 641-651. Springer Verlag.

Chorin, A. J. (1968) . Numerical solution of the Navier-Stokes equations. Math. Comp. , vol. 22 , pp 745-762 .

Chorin, A. J. and Marsden, J. E. (1979). A mathematical introduction to fluid mechanics. Springer Verlag.

Cuvelier, C. , Segal, A. and Steenhoven, A. A. (1986) . Finite Element Methods and Navier-Stokes Equations. D. Reidel Publishing Company.

Daly, B. J. and Harlow, F. H. (1970) . Transport equations in turbulence. The Physics of Fluids, vol. 13 , nr. 11 , pp 2634-2649.

Dean, R. G. (1965) . Stream function representation of nonlinear ocean waves. Jour. Geoph. Res. , vol. 70 , nr 18 , pp 4561-4572.

Dean, R. G. and Dalrymple, R. A. (1984) . Water wave mechanics for engineers and scientists. Prentice Hall.

Dijkstra, E. W. (1976) . A discipline of programming. Prentice Hall.

Dold, J. W. and Peregrine, D. H. (1984) . Steep unsteady water waves: an efficient computational scheme. Proc. Int. Conf. Coast. Eng. , pp 955-967.

Dold, J. W. and Peregrine, D. H. (1986) . An efficient boundary-integral method for steep unsteady water waves. In "Methods for Fluid Dynamics II", eds. K. W. Morton and M. J. Baines, OUP pp 671-679.

Dukowicz, J. K. and Ramshaw, J. D. (1979) . Tensor viscosity method for convection in numerical fluid dynamics. Jour. Comp. Phy. , vol. 32 , pp 71-79.

Favre, H. (1935) . Ondes de translation dans les canaux découverts. Paris, Dunod.

Fenton, J. D. (1972) . A ninth order solution for the solitary wave. Jour. Fluid Mech. , vol. 53 , pp 257-271.

Fenton, J. D. (1979) . A high-order cnoidal wave theory. Jour. Fluid Mech. , vol. 94 , pp 129-161.

Fenton, J. D. (1985) . A fifth-order Stokes theory for steady waves. Journal of Waterway, Port, Coastal and Ocean engineering, ASCE, vol. 111 nr. 2 , pp 216-234.

Fenton, J. D. and Rienecker, M. M. (1982) . A Fourier method for solving water-wave problems: application to solitary wave interactions. Jour. Fluid Mech. , vol. 118 , pp 411-443.

Fernholz, H. H. (1978) . External flows. In *Turbulence* , ed. P. Bradshaw . Springer Verlag.

Friedrichs, K. O. (1948) . On the derivation of the shallow water theory. Appendix to *The formation of breakers and bores*, by J. J. Stoker. Comm. Pure Appl. Math. , vol. 1 , pp 81-85.

Galvin, C. J. (1968) . Breaker type classification on three laboratory beaches. Jour. Geoph. Res. , vol. 73 , pp 3651-3659.

Galvin, C. J. (1972) . Wave breaking in shallow water. In *Waves on Beaches and Resulting sediment transport*, ed. R. E. Meyer, pp 413-456. New-York, Academic Press.

Goodrich, J. W. and Soh, W. Y. (1989) . Time-dependent incompressible Navier-Stokes equations: the finite difference Galerkin formulation and streamfunction algorithms. Jour. Comp. Phy. , vol. 84 , pp 207-241.

Gresho, P. M. (1990) . On the theory of semi-implicit projection methods for viscous incompressible flow and its implementation via a finite element method that also introduces a nearly consistent mass matrix. Part I : theory. Int. Jour. for Num. Meth. in Fluids, vol. 11 , pp 587-620.

Gresho, P. M. and Sani, R. L. (1987) . On pressure boundary conditions for the incompressible Navier-Stokes equations. Int. Jour. for Num. Meth. in Fluids, vol. 7 , pp 1111-1145.

Gresho, P. M. , Chan, S. T. , Lee, R. L. and Upson, C. D. (1984) . A modified finite element method for solving the time-dependent, incompressible Navier-Stokes equation. Part 1 : Theory . Int. Jour. for Num. Meth. in Fluids, vol. 4 , pp 557-598.

Gustafson, K. E. (1987) . Introduction to Partial Differential Equations and Hilbert Space Methods (second edition) .

Gustafson, K. , and Halasi, K. (1986) . Vortex dynamics of cavity flows. Jour. Comp. Phy. , vol. 64 , pp 279-319.

Gustafson, K. , and Halasi, K. (1987) . Cavity flow dynamics at higher Reynolds number and higher aspect ratio. Jour. Comp. Phy. , vol. 70 , pp 271-383.

Hanjalić, K. and Launder, B. E. (1976) . Contribution towards a Reynolds stress closure for low-Reynolds number turbulence. Jour. Fluid Mech. , vol. 74 , pp 593-610.

Harlow, F. H. and Nakayama, P. I. (1967) . Turbulence transport equations. The Physics of Fluids, vol. 10 , nr. 11 , pp 2323-2332.

Harlow, F. H. and Nakayama, P. I. (1968) . Transport of turbulence energy decay rate. Los Alamos Scientific Laboratory, report LA-3854.

Harlow, F. H. and Welch, J. E. (1965) . Numerical calculation of time-dependent viscous incompressible flow. The Physics of Fluids, vol. 8 , pp 2182-2189.

Hindmarsh, A. C. , Gresho, P. M. and Griffiths, D. F. (1984) . The stability of explicit Euler time-integration for certain finite-difference approximations of the multi-dimensional advection-diffusion equation. Int. Jour. for Num. Methods in Fluids, vol. 4 , pp 853-897 .

Hinze, J. O. (1975) . Turbulence (second edition). McGraw Hill Classic Textbook Reissue Series.

Hirt, C. W. (1968) . Heuristic stability theory for finite-difference equations. Jour. Comp. Phy. , vol. 2 , pp 339-355.

Hirt, C. W. (1969) . Computer studies of time-dependent turbulent flows. The Physics of Fluids supplement II, pp II-219 to II-227.

Hirt, C. W. and Cook, J. L. (1972) . Calculating three-dimensional flows around structures and over rough terrain. Jour. Comp. Phy. , vol. 10 , pp 324-340.

Hirt, C. W. , Nichols, B. D. and Romero, N. C. (1975) . SOLA : A numerical solution algorithm for transient fluid flows. Los Alamos Scientific Laboratory, report LA-5852.

Hirt, C. W. and Shannon, J. P. (1968) . Free-surface stress conditions for incompressible-flow calculations. Jour. Comp. Phy. , vol. 2 , pp 403-411.

Iribarren, C. R. and Nogales, C. (1949) . Protection des Ports, Section II, Comm. 4, 17th Int. Nav. Congress, Lisbon, pp 31-80.

Jansen, P. C. (1986) . Laboratory observations of the kinematics in the aerated region of breaking waves. Coastal Eng. , vol. 9 , pp 453-477.

Jones, W. P. and Launder, B. E. (1972) . The prediction of laminarization with a 2-equation model of turbulence. Int. Jour. Heat Mass Transfer, vol. 15 , pp 301-314.

Korteweg, D. J. & de Vries, G. (1895) . On the change of form of long waves advancing in a rectangular cannal and on a new type of long stationary wave. Phil. Mag. 39 (5), pp 422-443.

Laitone, E. V. (1960) . The second approximation to cnoidal and solitary waves. Jour. Fluid Mech. , vol. 9 , pp 430-443.

Lamb, H. (1932) . Hydrodynamics. Cambridge University Press.

Launder, B. E. (1979) . Stress transport closures–Into the third generation. In *Turbulent Shear Flows I* , eds. F. Durst et al. , Springer Verlag (1979) , pp 259-266.

Launder, B. E. (1989) . Second-moment closure and its use in modelling turbulent industrial flows. Int. Jour. for Num. Methods in Fluids , vol. 9 , pp 963-985.

Launder, B. E. and Spalding, D. B. (1972) . Mathematical Models of Turbulence. Academic Press.

Launder, B. E. , Reece, G. J. and Rodi, W. (1975) . Progress in the development of a Reynolds-stress turbulence closure. Jour. Fluid Mech. , vol. 68 , pp 537-566.

Lemoine, R. (1948) . Sur les ondes positives de translation dans les canaux et sur le ressault ondulé de faible amplitude. La Houille Blanche, Nr. 2 , Grenoble.

Leonard, A. (1974) . Energy cascade in large-eddy simulations of turbulent fluid flows. Adv. in Geophysics, 18A, pp 237-248.

Leonard, B. P. (1980) . Note on the von Neumann stability of the explicit FTCS convective diffusion equation. Appl. Modelling, vol. 4 , pp 401-402.

Lighthill, M. J. (1949) . A technique for rendering approximate solutions to physical problems uniformly valid. Phil. Mag. 40, pp 1179-1201.

Longuet-Higgins, M. S. (1978a) . The instabilities of gravity waves of finite amplitude in deep water. I. Super-Harmonics . Proc. Roy. Soc. Lond. A 360, pp 471-488.

Longuet-Higgins, M. S. (1978b) . The instabilities of gravity waves of finite amplitude in deep water. II. Sub-Harmonics . Proc. Roy. Soc. Lond. A 360, pp 489-505.

Longuet-Higgins, M. S. and Cokelet, E. D. (1976) . The deformation of steep surface waves on water. I. A numerical method of computation. Proc. Roy. Soc. Lond. A 350 , pp 1-26.

Longuet-Higgins, M. S. and Fenton, J. D. (1974) . On the mass, momentum, energy and circulation of a solitary wave. II. Proc. Roy. Soc. Lond. A 340, pp 471-493.

Madsen, P. A. and Svendsen, I. A. (1983) . Turbulent bores and hydraulic jumps. Jour. Fluid Mech. , vol. 129 , pp 1-25.

Maxworthy, T. (1976) . Experiments on collisions between solitary waves. Jour. Fluid Mech. , vol. 76 , pp 177-185.

McCowan, J. (1894) . On the highest wave of permanent type. Phil. Mag. 38 (5), pp 351-358.

Mei, C. C. (1983) . The applied dynamics of ocean surface waves. John Wiley & Sons.

Miles, J. W. (1980) . Solitary waves. Annual Review of Fluid Mechanics, Palo Alto, California, pp 11-43.

Miles, J. W. (1981) . The Korteweg-de Vries equation: a historical essay. Jour. Fluid Mech. , vol. 106 , pp 131-147.

Mizuguchi, M. (1986) . Experimental study on kinematics and dynamics of wave breaking. Proc. Int. Conf. of Coastal Eng. , pp 589-603.

Miyata, H. (1986) . Finite-difference simulation of breaking waves. Jour. Comp. Phy. , vol. 65 , pp 179-214 .

Morton, K. W. (1971) . Stability and convergence in fluid flow problems. Proc. Roy. Soc. Lond. A 323, pp 237-253.

Munk, W. H. (1949) . The solitary wave and its application to surf problems. Annals New York Academy of Science, vol. 51, pp 376-424.

Munk, W. H. and Wimbush, M. (1969) . A rule of the thumb for wave breaking over sloping beaches. Oceanology 6, pp 56-59.

Nadaoka, K. (1986) . A fundamental study on shoaling and velocity field structure of water waves in the nearshore zone. Tech. Rep. 36, Dept. Civil Eng. , Tokyo Inst. Techn. , Japan.

Nadaoka, K. and Kondoh, T. (1982) . Laboratory measurements of the velocity field structure in the surf zone by LDV. Coast. Eng. Japan, vol. 35 , pp 125-146.

Nichols, B. D. and Hirt, C. W. (1971) . Improved free-surface boundary conditions for numerical incompressible flow calculations. Jour. Comp. Phy. , vol. 8 , pp 434-448.

Nichols, B. D. and Hirt, C. W. (1975) . Methods for calculating multi-dimensional, transient free surface flows past bodies. Proc. of the 1st Intern. Conf. Num. Ship Hydrodynamics, Gaithersburg, Maryland, October 1975.

Nichols, B. D. , Hirt, C. W. and Hotchkiss, R. S. (1980) . SOLA-VOF : A solution algorithm for transient fluid flow with multiple free-boundaries. Los Alamos Scientific Laboratory, report LA-8355.

Oikawa, M. and Yajima, N. (1973) . Interactions of solitary waves — a perturbation approach to non-linear systems. Jour. Phys. Soc. Japan, vol. 34 , pp 1093-1099.

Peregrine, D. H. (1974) . Discussion. Jour. Hyd. Res. , vol. 1, pp 141-154.

Peregrine, D. H. (1983) . Breaking waves on beaches. Annual Review of Fluid Mechanics, Palo Alto, California, pp 149-178.

Peregrine, D. H. (1988) . Breaking water waves. Lectures given at the 109th International School of Physics "Enrico Fermi", organised by the Italian Physical Society. Villa Monastero, Varenna sul Lago di Como, Italy.

Peregrine, D. H. , Cokelet, E. D. and McIver, P. (1980) . The fluid mechanics of waves approaching breaking. Proc. Int. Conf. Coastal Eng. 17th, pp 512-528.

Peregrine, D. H. and Svendsen, I. A. (1978) . Spilling breakers, bores and hydraulic jumps. Proc. Int. Coastal Eng. Conf. 16th , Hamburg, vol. 1 , pp 540-550

Peyret, R. and Taylor, T. D. (1983) . Computational Methods for Fluid Flow. Springer Verlag.

Phillips, O. M. (1977) . The dynamics of the upper ocean (second edition). Cambridge University Press.

Poo, J. Y. and Ashgriz, N. (1989) . A computational method for determining curvatures. Jour. Comp. Phy. , vol. 84 , pp 483-491.

Rajaratnam, N. (1967) . Hydraulic jumps. Adv. Hydrosci. , vol. 4 , pp 197-280.

Resh, F. J. and Leutheusser, H. J. (1972) . Reynolds stress measurements in hydraulic jumps. Jour. Hyd. Res. , vol. 10 , pp 409-430.

Resh, F. J. , Leutheusser, H. J. and Coantic, M. (1976) . Étude de la structure cinématique et dynamique du ressault hydraulique. Jour. Hyd. Res. , vol. 14 , pp 293-319.

Reynolds, W. C. and Cebeci, T. (1978) . Calculation of Turbulent Flows. In *Turbulence* , ed. P. Bradshaw , Springer Verlag.

Richtmeyer, R. D. and Morton, K. W. (1967) . Difference methods for initial-value problems (second edition). Interscience Publishers, J. Wiley & Sons , New York.

Rienecker, M. M. and Fenton, J. D. (1981) . A Fourier approximation method for steady water waves. Jour. Fluid Mech. , vol. 104 , pp 119-137.

Roache, P. (1972) . Computational Fluid Dynamics. Hermosa Publishers, Albuquerque, U. S. A. .

Rodi, W. (1980) . Turbulence Models and Their Application in Hydraulics–A State-of-the-art review. I. A. H. R. publication.

Saffman, P. G. (1981) . Dynamics of vorticity. Jour. Fluid Mech. , vol. 106 , pp 49-58 .

Schwartz, L. W. (1974) . Computer extension and analytic continuation of Stokes' expansion for gravity waves. Jour. Fluid Mech. , vol. 62 , pp 553-578.

Schwartz, L. W. and Fenton, J. D. (1982) . Strongly nonlinear waves. Annual Review of Fluid Mechanics, Palo Alto, California, pp 39-60.

Smith, G. D. (1985) . Numerical solution of partial differential equations: Finite difference methods. Clarendon Press-Oxford .

Smith, I. M. (1982) . Programming the finite element method, with application to geomechanics. J. Wiley.

Smolarkiewicz, P. K. (1984) . The fully multidimensional positive definite advection transport algorithm with small implicit diffusion. Jour. Comp. Phy. , vol. 54 , pp 325-362.

Stiassnie, M. and Peregrine, D. H. (1980) . Shoaling of finite-amplitude surface waves on water of slowly-varying depth. Jour. Fluid Mech. , vol. 97 , part 4, pp 783-805.

Stokes, G. G. (1847) . On the theory of oscillatory waves. Trans. Cambridge Philos. Soc. , vol.8 , pp 441-455.

Sumer, B. M. , Jensen, B. L. and Fredsöe, J. (1986) . Turbulence in oscillatory boundary layers. In *Advances in Turbulence* , eds. G. Comte-Belot and J. Mathieu, pp 556-567. Springer Verlag .

Svendsen, I. A. (1987) . Analysis of surf-zone turbulence. Jour. Geoph. Res. , vol. 92, pp 5115-5124 .

Svendsen, I. A. and Madsen, P. A. (1984) . A turbulent bore on a beach. Jour. Fluid Mech. , vol. 148 , pp 73-96.

Tanaka, M. (1986) . The stability of solitary waves. The Physics of Fluids, vol. 29, pp 650-655.

Tanaka, M. , Dold, J. W. , Lewy, M. and Peregrine, D. H. (1987) . Instability and breaking of a solitary wave. Jour. Fluid Mech. , vol. 185 , pp 235-248.

Teles da Silva, A. F. and Peregrine, D. H. (1990) . Nonsteady computations of undular and breaking bores. Int. Conf. Coastal Eng. 22th, Delft (to appear).

Temam, R. (1983) . Navier-Stokes equations and nonlinear functional analysis. SIAM, Philadelphia.

Tennekes, H. and Lumley, J. L. (1972) . A First Course in Turbulence. MIT press.

Townsend A. A. (1976) . The structure of turbulent shear flow (second edition). Cambridge University Press.

Ursell, F. (1953) . The long-wave paradox in the theory of gravity waves. Proc. Camb. Phil. Soc. , vol. 49 , pp 685-694.

Viecelli, J. A. (1969) . A method for including arbitrary external boundaries in the MAC incompressible fluid computing technique. Jour. Comp. Phy. , vol. 4 , pp 543-551.

Viecelli, J. A. (1971) . A computing method for incompressible flows bounded by moving walls. Jour. Comp. Phy. , vol. 8 , pp 119-143.

Vinje, T. and Brevig, P. (1981) . Numerical simulation of breaking waves. Adv. Water Resources 4, pp 77-82.

Weidman, P. D. and Maxworthy, T. (1978) . Experiments on strong interactions between solitary waves. Jour. Fluid Mech. , vol. 85 , pp 417-431.

Welch, J. E. , Harlow, F. , Shannon , J. P. and Daly, B. J. (1966) . The MAC method: A computing technique for solving viscous, incompressible, transient fluid flow problems involving free-surfaces. Los Alamos Scientific Laboratory, report LA-3425.

Welland, R. (1986) . Structured programming. Pitman computer handbooks, U. K. .

Whitham, G. B. (1974) . Linear and nonlinear waves. Interscience Publishers, J. Wiley & Sons, New York.

Zalesak, S. T. (1979) . Fully multidimensional flux corrected transport. Jour. Comp. Phy. , vol. 31 , pp 335-362.

Lecture Notes in Engineering

Edited by C.A. Brebbia and S.A. Orszag

Vol. 40: R. Borghi, S. N. B. Murhty (Eds.)
Turbulent Reactive Flows
VIII, 950 pages. 1989

Vol. 41: W. J. Lick
Difference Equations
from Differential Equations
X, 282 pages. 1989

Vol. 42: H. A. Eschenauer, G. Thierauf (Eds.)
Discretization Methods
and Structural Optimization –
Procedures and Applications
Proceedings of a GAMM-Seminar
October 5-7, 1988, Siegen, FRG
XV, 360 pages. 1989

Vol. 43: C. C. Chao, S. A. Orszag, W. Shyy (Eds.)
Recent Advances in Computational
Fluid Dynamics
Proceedings of the US/ROC (Taiwan) Joint
Workshop in Recent Advances in
Computational Fluid Dynamics
V, 529 pages. 1989

Vol. 44: R. S. Edgar
Field Analysis and
Potential Theory
XII, 696 pages. 1989

Vol. 45: M. Gad-el-Hak (Ed.)
Advances in Fluid Mechanics
Measurements
VII, 606 pages. 1989

Vol. 46: M. Gad-el-Hak (Ed.)
Frontiers in Experimental
Fluid Mechanics
VI, 532 pages. 1989

Vol. 47: H. W. Bergmann (Ed.)
Optimization: Methods and Applications,
Possibilities and Limitations
Proceedings of an International Seminar
Organized by Deutsche Forschungsanstalt für
Luft- und Raumfahrt (DLR), Bonn, June 1989
IV, 155 pages. 1989

Vol. 48: P. Thoft-Christensen (Ed.)
Reliability and Optimization
of Structural Systems '88
Proceedings of the 2nd IFIP WG 7.5 Conference
London, UK, September 26-28, 1988
VII, 434 pages. 1989

Vol. 49: J. P. Boyd
Chebyshev & Fourier Spectral Methods
XVI, 798 pages. 1989

Vol. 50: L. Chibani
Optimum Design of Structures
VIII, 154 pages. 1989

Vol. 51: G. Karami
A Boundary Element Method for
Two-Dimensional Contact Problems
VII, 243 pages. 1989

Vol. 52: Y. S. Jiang
Slope Analysis Using
Boundary Elements
IV, 176 pages. 1989

Vol. 53: A. S. Jovanovic,
K. F. Kussmaul, A. C. Lucia,
P. P. Bonissone (Eds.)
Expert Systems in Structural
Safety Assessment
X, 493 pages. 1989

Vol. 54: T. J. Mueller (Ed.)
Low Reynolds Number
Aerodynamics
V, 446 pages. 1989

Vol. 55: K. Kitagawa
Boundary Element Analysis
of Viscous Flow
VII, 136 pages. 1990

Vol. 56: A. A. Aldama
Filtering Techniques for
Turbulent Flow Simulation
VIII, 397 pages. 1990

Vol. 57: M. G. Donley, P. D. Spanos
Dynamic Analysis of Non-Linear
Structures by the Method of
Statistical Quadratization
VII, 186 pages. 1990

Vol. 58: S. Naomis, P. C. M. Lau
Computational Tensor Analysis
of Shell Structures
XII, 304 pages. 1990

GPSR Compliance

The European Union's (EU) General Product Safety Regulation (GPSR) is a set of rules that requires consumer products to be safe and our obligations to ensure this.

If you have any concerns about our products, you can contact us on ProductSafety@springernature.com

In case Publisher is established outside the EU, the EU authorized representative is:

Springer Nature Customer Service Center GmbH
Europaplatz 3
69115 Heidelberg, Germany

Batch number: 10704771

Printed by Printforce, the Netherlands